THE LIVING SOIL
AND
THE HAUGHLEY EXPERIMENT

E. B. BALFOUR

THE LIVING SOIL

and

THE HAUGHLEY EXPERIMENT

FABER AND FABER
3 Queen Square
London

Part One was first published as The Living Soil *in 1943*
by Faber and Faber Limited
3 Queen Square London WC1
New and revised edition 1975
Printed in Great Britain by
Latimer Trend & Company Ltd Plymouth

ISBN 0 571 10713 3

Contents

PART ONE
THE LIVING SOIL

PART TWO

THE STORY OF THE HAUGHLEY EXPERIMENT

(*with the collaboration of*
Dr. R. F. Milton, BSC, PHD, FRIC, FIBIOL, FIFST)

PART ONE

THE LIVING SOIL

Evidence of the importance
to human health of
soil vitality

Foreword to the New and Revised Edition

A great deal has happened since the data presented in this book were first collected and assembled. Research during the two decades that have passed since the publication of the 1948 edition has fundamentally altered the current concepts at that time, both of nutrition and of the nature of the soil. Yet the new knowledge, far from invalidating the work and ideas of the early pioneers—outlined in the following chapters—has served to confirm the importance of their work and to show how far-seeing they were. For while much of the early record, reprinted as Part I of the present volume, now requires amplification, I have found nothing that requires contradiction. It therefore remains a good background from which to consider the history of more recent developments.

During the late 1930s, while I was collecting and collating the evidence for this book, two things impressed me more and more: the unparalleled importance of the new field opening up before me, and the paramount need for new scientific research—new because it must be based on a new concept, and break away from the fetters of long-established fragmentary techniques.

The concept I refer to—quite new at that time—is that health, whether of soil, plant, animal or man, is one and indivisible. It consists of 'mutual synthesis between organism and environment'.[1] Inherent in such a definition is the idea that health is not a state but a dynamic process. Any study concerning it must therefore be qualitative, not quantitative. For you cannot weigh, measure, or analyse a process, but it is sometimes possible to follow and observe its course. The early pioneers believed that its course is identical with the flow of the nutrition cycle, and that to promote it one must, therefore, keep open all the living channels of this flow, though no one yet knows what they all are, or even the true nature of the flow itself. That land

[1] I. H. Pearse and L. H. Crocker, *The Peckham Experiment*, Allen & Unwin, London, 1943; reprinted in paperback.

is a great storehouse for it, however, seems clear. What then is land? Let me give the late Aldo Leopold's definition:[1] 'Land . . . is not merely soil; it is a fountain of energy flowing through a circuit of soils, plants and animals. Food chains are the living channels which conduct energy upward; death and decay return it to the soil.' Soil fertility he defined as 'the capacity of soil to receive, store and transmit energy'.

The concept that the nutrition cycle is not merely a transfer of nutrient materials from one form of life to another, but also a circuit of energy, though even now not universally accepted, is no longer considered revolutionary, and under the name of ecology has become a respectable subject for research. Part II of this book is devoted to a description of the earliest investigation of this kind, which even today remains unique in its way, for research into 'wholes' is notoriously difficult, and any study of this particular form of 'energy' (which distinguishes the living from the dead) is fraught with difficulties because, among other things, it is not yet capable of measurement by *any* recognized science. If, as seems probable, it can be transmitted only by life, the difference in resulting plant growth following the application of, for example, a living compost and the ash of the same compost after incineration, can readily be understood. The fallacy of basing the supposed needs of plants on the composition of their ash also becomes apparent. Something chemically the same may be very different as a conductor of living energy.

Even during the first decade following publication of *The Living Soil* advances in biological research had demonstrated the total inadequacy of the science of chemistry to explain the complexity of living processes—this resulted in discrediting much of the dogma which had governed soil science up to that time.

Notable among the discoveries between 1948 and 1958 were the following:

(a) That plants are able to absorb through their roots unbroken molecules of quite complicated organic compounds, whether man-made and poisonous, like DDT, or the beneficial products of micro-biological soil life. This discovery has swept away forever the dogma that all plant food must be reduced to a solution of simple salt before it can be utilized by plants. Thus the bottom has dropped out of the argument that there is no difference between a plant nutrient contained in organic materials and the same nutrient in inorganic chemical form.

(b and c) The work which has been, and is still being, done on trace

[1] A. Leopold, *Sand County Almanac*, OUP (NY), 1969.

minerals and proteins, which has revealed the extremely complex nature of the latter, has resulted in the disappearance of another previously firmly held dogma, namely, that a high protein content in fodder (hay, grass, silage, etc.) is a measure of its value as a foodstuff. It is now known that what matters is the *quality* of the protein, and that depends (according to present knowledge) on the range of amino acids composing it. Plant protein may, or may not, contain certain amino acids which are essential to animal nutrition. Whether they do or not depends largely on the soil conditions in which the plant is grown, for the enzyme systems which control cell metabolism are dependent on trace minerals, and the availability of these to the plant is dependent on soil micro-organisms.

Even a slight deficiency in one of a given range of trace minerals has been shown, in recent research, to result, not only in the absence in the plant protein of such amino acids as triptophane and lysine (two of the musts for animal health) but can also lead to the plants being attacked by aphis, while neighbouring plants, not suffering from any deficiency, remain free from attack.[1]

This throws new light on the often reported pest resistance of compost-grown crops. I have myself seen two lettuce crops divided only by a foot-wide strip of grass, one belonging to an orthodox holding, the other to an organic holding. The first was nearly destroyed by aphis, despite dustings; on the other I failed to find a single insect.

It now seems reasonably well established that on this interplay between soil micro-organisms and the trace elements may depend, not only the health and disease resistance of plants, but the range of amino acids contained in their protein. Also that on this quality of plant protein depends the health and disease resistance of the animal (and probably the human being) who consumes it. Hence, since soil micro-organisms play such an important part in this chain reaction, the biological importance of humus (their food and habitat) becomes plain for all to see.

(d) There have been other discoveries concerning the soil population. These have not resulted from one line of research, but many. For example, Dr. Swaby's thorough investigation into the causation of good crumb structure in soil proved that the most important factor was the action of soil fungi *while living*.[2] Dr. Dhar's work in India,

[1] See *Mother Earth*, April 1954, or *9,600 miles in a Station Wagon* re research by Dr. William Albrecht (both obtainable from the Soil Association, see p. 370).

[2] See *Mother Earth*, April 1961, re Dr. Swaby.

Sweden and elsewhere has demonstrated that when powdered rock phosphate or finely ground basic slag is added to crude organic matter, such as straw, before it is ploughed into the soil, not only does this prevent subsequent nitrogen starvation in the soil with which it is incorporated, but that it leads to a higher nitrogen content in the soil than arises from the addition of artificial nitrogen; allegedly because phosphate plus organic matter stimulates the free-nitrogen-fixing organisms to a greater degree than artificial nitrogen assists the cellulose-decomposing ones.[1]

Dr. Duddington's work with predaceous fungi has shown inoculation of compost, or FYM, with specially developed cultures of these fungi promises to be a most effective way of dealing with eel worm infestation.[2]

Besides such specialized discoveries, and those already referred to above, modern biological studies of soil have revealed, and continue to reveal how little we yet know of its extraordinary complexity, and how utterly dependent plant life is on the micro-organic life in the soil.[3]

Small wonder then that the orthodox agricultural teaching of a few years ago has received a severe shaking.

The history of the advance in knowledge concerning human nutrition during the past two decades follows the same pattern—that the more that is discovered the more complex it becomes.

So at last we see the apparently separate trails blazed by McCarrison and Howard (described in Part I) coming together. Research scientists in nutrition and in agriculture are beginning to recognize that they are fellow students working in the same field of life forces.[4]

An ever-increasing number of authorities nowadays would agree with Professor R. Lindsay Robb's statement:

'Since human health and soil fertility can both be expressed as their capacity to absorb, store and release energy over a continuing

[1] N. Dhar, *World Food Crisis and Land Fertility Improvement*, University of Calcutta, 1972.

[2] C. L. Duddington, *Friendly Fungi*, Faber & Faber, London, 1957.

[3] Sir E. John Russell, *The World of the Soil*, Fontana, London, 1970. See also Part II, p. 208.

[4] Very recent research carried out by Dr. Cleve Backster and other scientists (some of them Russian) into sensitivity and apparent 'extra-sensory perception' in plants is producing striking evidence indicating that every living organism is linked to every other living organism through a, so far, unidentified energy field. This evidence to date has been admirably collected and brought together in Peter Tompkins and Christopher Bird, *The Secret Life of Plants*, Harper & Row, New York, 1973 and Allen Lane, London, 1974.

period, it seems highly probable that the same foods may vary in health value in accordance with the level of fertility of the soils on which they were grown. We measure food production in terms of yield, or quantity per unit area, but the real measure is not mere quantity but total human health value of the crop produced. From this it follows naturally that there is a new world to explore in agro-medical-ecological research, because, in the final analysis, the function of land is to maintain human health, and from the handful of soil on the one hand to the loaf of bread on the other this should never be lost sight of in what goes on between.'[1]

I must end with a few explanatory notes concerning the reprint of the first seven chapters of *The Living Soil* (first published in 1943) which now forms Part I of the present book.

It should be regarded as source-material and as such of historical value to the organic movement.

I have edited it slightly here and there and made a few cuts, but I have made no attempt to bring it up to date, indeed that would not have been possible. Where it appears desirable, however, to draw special attention to the time of writing, or of the event described, I have put a short paragraph of comment at the head of the chapter concerned, in square brackets. Any additional footnote, or new note in the text, I have also put in square brackets. The original Technical References have been pruned and those remaining have been placed at the end of the chapters to which they relate.

I have left in the greater part of the short original introduction; partly because it is still valid, but also because it contains the acknowledgements.

In an attempt to keep the price of the book down to a reasonable level it has been found necessary to omit the illustrations in this edition.

I donated the royalties of the earlier editions towards the support of the Haughley Experiment. Since this has now come to an end, royalties on the present book will be given to the Soil Association.

That is all that needs to be said here. Part II, *The Story of the Haughley Experiment*, has its own introductory material.

E.B.B. 1974

[1] *Man and the Soil, Mountain Top to City Pavement*. Second Sanderson-Wells Lecture, 1957. Delivered at London University Senate House on 28 May 1957.

From the Introduction to the 1943, 1946 and 1948 Editions

In this book I have attempted something which my friends tell me cannot be done. I have tried to write for both the specialist and the layman. If I have failed, if what I have written proves to be unintelligible and boring to the layman, and at the same time trite and superficial to the specialist, I shall be unable to plead that I was not warned, but my only alternative was not to write at all—a depressing one for the would-be author suffering from the belief that he has something important to say. At any rate I have refused to accept such defeatism, preferring to attempt the supposedly impossible. For there is a good reason why this book could not be written for a limited section of the reading public.

My subject is food, which concerns everyone; it is health, which concerns everyone; it is the soil, which concerns everyone—even if they do not realize it—and it is the history of certain recent scientific research linking these three vital subjects.

Now since it concerns health, and food, and soil, and science, it concerns the legislator, politician, voter, tax- and rate-payer; farmer, gardener, veterinary surgeon, doctor; sanitary inspector, public health authority, school teacher, priest; 'Tinker, Tailor, Soldier, Sailor . . .'—in fact, the Citizen. So there was no third choice; I had to write for everyone or not at all.

I have, however, tried to arrive at a compromise, whereby what I have written will, I hope, be acceptable both to the citizen who is also a specialist, and to the citizen who—perhaps because of our modern tendency to urbanize everything—goes by the name of 'the man in the street'.

I have avoided as much as possible the inclusion of too much technical detail in the main text, but have included [at the end of chapters in Part I] additional technical data on some of the subjects dealt with. To these the specialist can refer should he want more precise detail concerning the evidence presented.

As it has been impossible to avoid entirely the use of technical and scientific terms, particularly in Chapters 4 and 5, I have provided a glossary for the benefit of those to whom these terms may be unfamiliar.

It is my earnest hope that by these means I shall have confounded the pessimists, for it is quite impossible to overestimate the importance of the subject. That this is becoming increasingly realized, is shown by the spate of publications that have appeared recently concerning it. I make no apology for adding to their number. There cannot be too many, and no two people approach a subject as complex as this from exactly the same angle.

The reader will find that much of this book consists of quotations. I make no apology for this either. The information required to present the picture as a whole—and the subject really amounts to the ecology of life—is so scattered, that the ordinary citizen could not possibly be expected to hunt it all up for himself, even if he knew where to look. One of my objects in undertaking this book has been to bring together, within the covers of a single volume, a summary of certain recent scientific research on nutrition and on soil fertility, including results that have been obtained in many different parts of the world by practical application of the principles involved.

The cumulative evidence which these results provide of the interrelationship between soil vitality and the health of plants, animals and man, is of so important a nature, and of such far-reaching implication, that it is high time the general public were given an opportunity to study it, and to form its own judgement thereon: and here, in order to avoid any misunderstanding, I would like to insert a note on terminology.

First. *Soil fertility.* This term is used throughout this book to denote vitality. It is not used as a synonym for productivity. Increased productivity can, and usually does, result from increased vitality. But for a limited period it can also be induced at the *expense* of vitality. In the former case the increase represents increased income, in the latter it represents reduced capital.

Second. *Artificial (or Chemical) Fertilizers.* These words as used throughout this book refer to manufactured synthetic inorganic chemicals, or more simply still, fertilizers not derived from living, or once living, matter. 'All artificial things are invented, and all are "made" by man. All natural entities are born, and all are grown by Nature.'[1] Quite definitely the term artificial, used in this sense, does not include such organic products as chalk, guano, pure bone or fish meal, etc. Obviously the addition to the soil of animal residues such as these cannot detract from its vital quality, which is the point at issue. The basis of the whole theme is an ecological one, centring round the mutual relationship of the organism and its environment. Considerable light is thrown on the importance to health of a living environment, by the evidence recently published in *The Peckham Experiment*, by Pearse and Crocker.[2]

I am very much aware of the magnitude of the task I have undertaken in attempting to present so vast a theme, and of the impossibility of covering the subject adequately in the space of a single volume. All I can hope to do is to convince the reader of the importance of the issues involved and to whet his appetite for more information. If I succeed in achieving so much, the writing of this book will have been worth while.

I need hardly say that I could not have attempted to write it unaided. The number of people to whom I am indebted for assistance of one kind or another is so large that it is impossible to list them all individually, but as everyone's help was, without exception, of the most generous and disinterested kind, any, whose names I have failed to acknowledge here, will know that they are nevertheless included in my thanks. Special mention must, however, be made of the following:

Dr. L. J. Picton, for invaluable help with the pamphlet of which this book is really an enlargement. Dr. M. C. Rayner and Professor Neilson-Jones, not only for the contribution of vital material, but also for critical advice, given whenever asked, without stint, and at considerable expenditure of time and trouble. But for the generous way in which these two experts put their knowledge at my disposal, the evidence which I have been able to present in this book would have been very much less complete. Mrs. Ysabel Daldy, for giving me invaluable information concerning New Zealand. The personnel of the Association for Planning and Regional Reconstruction, for the trouble taken in ferreting out tiresome statistics for me. Then there is the long list of those who have supplied me with material and information, or have allowed me to quote from their writings. These include, in addition to all the above, Captain R. G. M. Wilson; the County Palatine of Chester Local Medical and Panel Committees; the

[1] G. Scott Williamson, M.D.
[2] Allen & Unwin, London, 1943.

Earl of Portsmouth; Dr. G. T. Wrench; Sir Albert Howard; Messrs. Jacks and Whyte; Dr. K. E. Barlow; Lord Northbourne; Mr. C. C. J. Bullough; Viscount Bledisloe, and many friends who follow the honourable profession of tilling the soil.

My thanks are also due to the following publishers and editors, without whose co-operation the permission of the authors concerned would have been inoperative. Daniel; Dent; Faber and Faber; Oxford University Press; Cambridge University Press; the editors of *Forestry* (the Journal of the Society of the Foresters of Great Britain); the editor of the *Journal of Agricultural Science*; the editors of *Nature*; the editor of the *Empire Cotton Growing Review*; the editor of the *Compost News Letter*.

I am also indebted to the following for some of my facts: PEP Biological Reviews; Dr. Charles Drechsler; the Winsford UDC; Mr. F. H. Billington; Messrs. Baker and Martin; Association for Planning and Regional Reconstruction; and various journals and newspapers.

Haughley { *February 1943*, *December 1946*, *January 1948* }

E.B.B.

'Healthy citizens are the greatest asset any country can have.'

WINSTON CHURCHILL

World Broadcast, 22 March 1943

I

Preliminary Survey

[At the time this chapter was written, the threat to survival posed by soil erosion appeared, to those in the know, to be the dominant environmental problem. Today, that of pollution seems to have superseded it in importance, and may well prove to be the most intractable of the two. Nevertheless, soil erosion, despite all the progress that has been made in conservation education and practices, is still with us and should not be forgotten.

So far as erosion is concerned some of the trends described below have begun to be reversed. In both Kenya and Australia, for example, much land believed destroyed for ever has been restored and is growing good crops. Looked at on a world scale, however, the facts are still much as described in this chapter.]

'If mankind cannot devise and enforce ways of dealing with the earth, which will preserve the source of life, we must look forward to a time—remote it may be, yet clearly discernible—when our kind, having wasted its great inheritance, will fade from the earth because of the ruin it has accomplished.'

Those words were written by Professor N. S. Shaler of Harvard University in 1896.[1] Today that time of which he spoke can no longer be termed remote. Does this sound like a wild and exaggerated statement? Those who have made a study of soil erosion throughout the world know only too well the truth of it. Read Lord Northbourne's summary of what is happening in our world today.

'Erosion is always going on, even on some soils which are in a good state of fertility. It begins to matter when the rate of erosion exceeds the rate at which life can invade the mineral rock underlying the soil and convert it into soil. That rate is variable, but always very slow; of the order of one inch in 500 to 1,000 years. It matters desperately when the rate of erosion mounts up to very much higher rates than that, as it often does. A high erosion rate in a few limited areas would be regrettable; but when it covers whole continents, as it does today, the fate of the world, and of man with it, must be hanging in the

[1] *National Geographic Magazine* of that year.

balance. It is very difficult to form a scientifically reliable estimate of the real extent of erosion; but there is enough evidence to show that its incidence is world-wide and severe. No country is wholly exempt, but the big continental areas are generally the most seriously affected. There is a fairly large and growing amount of literature on the subject. Probably the most comprehensive survey is *The Rape of the Earth* (1939) by G. V. Jacks and R. O. Whyte.

'The country which has received most attention in this connection is the USA, and deservedly so, for America as usual is out for records. Alarming statistics can be quoted endlessly. On 56·4 per cent of the land surface of the USA, a quarter or more of the soil has been lost. The total loss of fertility has been estimated at 30–50 per cent of the total originally available. The amount of soil annually reaching the sea is between 500 and 1,000 million tons, representing 2,000 million dollars' worth of plant food, or twenty-one times the amount annually removed in crops; but that represents only a fraction of the total damage. Fifteen million acres have been totally destroyed, but this is "an insignificant part of the story, for it is the less violent forms of wastage—sheet erosion—which is doing the bulk of the damage to the land". The Missouri basin has lost an average of 7 inches of top soil in twenty-four years. (Professor Chamberlin has estimated the mean rate of soil formation as only 1 inch in 10,000 years.) In California and elsewhere the new deserts are called "dustbowls". The biggest one has advanced in places as much as forty miles in one year, destroying 2,500 farms. Efforts to stop it by tree planting, etc., have failed. The grazing lands of the West are not exempt, for over-grazing and fires have removed the natural cover. Obviously every other problem with which America is faced sinks into insignificance in comparison with this one. It is already too late to do more than save something from the wreck.

'Much the same is true of many other countries. Australia is probably going faster than America, but has only been under "civilized" influence for one-third of the period. Overstocking and unsound cultural methods are the chief troubles there; there is much gully erosion. The wheat lands of New South Wales are said to be getting visibly worse each year.

'In Africa, the Sahara desert is moving southward at a mean rate of over half a mile a year, the Turkana desert eastward at six or seven miles a year. But the whole continent is suffering from erosion in every known form, the extension of deserts and the creation of new ones. It is well known that Kenya is rapidly becoming infertile and is beginning to suffer from locusts. This is no new phenomenon in Africa, for it is known that the northern Sahara was once the granary of Rome, and it is believed that in Roman times the Congo forest reached nearly to Khartoum, from which it is now separated by 1,500 miles of desert or semi-desert. Erosion is not new, perhaps, but the whole process has been enormously accelerated in the last few years.

'China presents remarkable contrasts between the best and the worst. Some of the best cultivation in the world is practised over extensive areas (the reader is referred to Professor King's classic, *Farmers of Forty Centuries*), yet over areas far more extensive the worst types of erosion prevail. The well-cultivated areas are by no means immune from its effects. Chinese industry and care for the land is unequalled, yet it cannot prevent the periodical destruction caused by the Yellow River floods. This river alone carries down 2,500 million tons of eroded soil a year, equal to one foot on 2,000 square miles. Its bed gets silted up between embankments, which must be consistently raised till its bed is well above the surrounding land. Nothing can save that land or its people when an embankment bursts. All that is largely because fuel is scarce in China and the hills have been denuded to provide it. In this way what was once the hunting-ground of Genghis Khan has been turned into the Gobi desert.

'In Russia erosion on a serious scale has been going on for a long time, though no reliable estimate of its extent seems to be available. The important feature there is that it has been very greatly accelerated by the much-vaunted large-scale mechanical cultivation which has been introduced in recent years. Russia, with her vast areas of low rainfall and high winds, may be well on the way to rivalling America as a record-holder in rapidity and extent of desert formation.

'In Canada, South America, and India the same tendencies are observable in varying incidence, with, so far as can be seen, no compensating tendency towards increase of fertility. Increase of production must not be confused with increase of fertility. Increased production for human use can be and usually is secured by cashing in on existing fertility and using it up, with the disastrous effects described.

'Most of the land area of the world has now been mentioned, with the exception of Europe. In Europe actual erosion has not occurred on a spectacular scale. Nevertheless it is known, for instance, that in Greece the hilltops were once forests and the slopes were covered with soil and grass. They are not so today. It is also known that there has been a progressive desiccation of the soil in the whole Mediterranean region during at least the past 300 years. Terraced cultivation used to be more prevalent and more carefully maintained than is the case today, and forests have been destroyed for fuel. There is reason to suppose that the washing-down of hillsides is proceeding in places at a dangerous rate; for instance, in Savoy at least 100,000 acres of good land have been spoilt by coarse silt deposited on them during floods. In England the blowing away of certain of the light fen soils has attracted attention recently.

'But serious erosion is only the culminating stage of a process of which the initial stage is usually loss of fertility of one kind or another. Loss of fertility is vastly more extensive in its incidence than erosion. A picture of the extent of erosion gives merely an indication of the much greater incidence of loss of fertility. If erosion represents the death of the soil, and is as extensive as it appears to be, how much land is partway towards death? The question would be pertinent if the rate of erosion were steady. But it is not steady, it is increasing very rapidly almost all over the world. *Probably more soil has been lost since 1914 than in the whole previous history of the world.*[1] This is not a natural phenomenon in the ordinary sense of the word. There cannot be any doubt that so far as the modern growth of deserts is concerned it is not nature but man who is the desert maker. It is not unlikely that most of the great deserts of the world are of his making. When we consider how he sets about it now in conjunction with the fact that traces of high civilization are found in many areas now desert, the probability of his past guilt becomes greater. And the exhaustion of the fertility of the soil is no new thing, nor is the temptation to practise it for immediate gain. The new feature in the situation is that man has recently enormously extended his physical powers by the use of mechanical devices. One man can now do what used to be the work of dozens or even hundreds, and can do it faster.

'Natural erosion is either very slow, or, where it is relatively rapid, it occurs only under special conditions, such as are found on high and barren mountains. Nor does a high rate of erosion necessarily accompany cultivation. Some areas, for instance parts of China, have been highly cultivated for at least 4,000 years without loss of fertility, and probably with an accretion of fertile soil rather than with a loss of soil.

'Man sets about his desert-making in various ways. He alters the texture of the soil by using up humus and failing to replace it—by failing to feed the soil with organic matter; livestock are the great converters of otherwise unwanted organic matter to a form in which it can be used by plants. Stockless farming, understocking, burning

[1] Italics mine (author).

straw, etc., are all cases of failure to observe the "Rule of Return" which is the essence of farming. Only by faithfully returning to the soil in due course everything that has come from it can fertility be made permanent and the earth be made to yield a genuine increase.'[1]

Some of the repercussions of the rapidly extending area of man-made deserts will be considered in a later chapter. At the moment the sentence in this quotation to which I want specially to draw your attention is: 'If erosion represents the death of the soil . . . how much land is partway towards death?' This phrase, 'the death of the soil' is no figure of speech. Soil is a substance teeming with life. If this life is killed, the soil quite literally dies. It is the living organisms in soil, and the products resulting from their activities, that differentiate soil from subsoil.

Subsoil is derived from the crumbling surface of the rock which forms the earth's crust. It is therefore classified as mineral, and measured in terms of inorganic chemistry. The mistake has been to extend this conception to top soil, for by the time subsoil becomes soil it is no longer wholly inorganic. Soil is a mixture of the disintegrating mineral rock, and humus, with its population of micro-organisms.[2] The two are inseparably intermingled, and growth and reproduction cannot long be maintained under natural conditions in the absence of humus, for besides the growth-promoting substances which it contains, it is the humus which gives to soil its texture, its stability, and much of its capacity for retaining moisture.

Humus is 'a product of the decomposition of animal and vegetable residues brought about through the agency of micro-organisms' (Waksman) but it is far from dead in the sense of having returned to the inorganic world from which all life is commonly supposed to have originated. It is still organic matter, in the transition stage between one form of life and another. Once the inorganic passes into the organic, and this is a constant process, it is subject to continual change within the organic cycle, the variety in the forms of life through which it may pass being almost endless.

In our modern world, which is largely ruled by chemistry, we have tended to overlook this continuity of the living principle in nature. Chemists have discovered that we, and all other living things, consist of a few chemicals and a lot of water, but their methods are incapable

[1] Lord Northbourne, *Look to the Land*, Dent, London, 1930.

[2] The number of micro-organisms in soil has been estimated to reach tens of millions to the *saltspoon full*! F. H. Billington, *Compost*, Faber & Faber London, 1942.

of revealing the essential nature of the most important ingredient of all, because that ingredient does not survive the tests necessary to determine the others. I refer of course to the ingredient of life itself, which permeates each individual cell of all the countless millions that go to make up the plant or animal body.

Our over-reliance on the chemist, and our readiness to accept his negative outlook on life, have led to a host of troubles, not least of which has been our attitude to the soil, and our habit of thinking of birth and death—in so far, at any rate, as the physical world is concerned—as a beginning and an end, instead of as merely two phases of a continuous process, of which other phases are growth, reproduction and decay, all equally important.

This ever-recurring cycle of birth, growth, reproduction, death, decay, decay passing once more into birth, is often called the Wheel of Life. Such similes irritate some people and are helpful to others; I have rather a weakness for them myself, though they can often be misleading. In this case, however, I would choose to liken life to a continuous cable rather than to a wheel, the organic cycle represented by one of its strands, and each phase of it, birth, death, etc., as one of the threads in that strand. For there are other strands in life. The inorganic cycle is one; spiritual values are another. These various strands run parallel to each other, but are nevertheless interwoven at every point just as are the separate strands of a cable, so that if any thread of any of these strands is broken the whole cable is thereby weakened. This book is mainly concerned with the strand representing the organic cycle, for it is man's ignorant interference with this cycle which has produced the effects now threatening his very existence. By ignoring the law of return man has weakened the thread called decay. It is perilously near breaking. If we let it break beyond repair we face starvation.

'Out of the earth are we and the plants and animals that feed us created and made, and to the earth we must return the things whereof we are made if it is to yield again foods of a quality suited to our needs.'[1]

Evidence is steadily mounting to suggest that our failure to conform to this law of return is already producing, even in this country, the first signs of a dying soil, and the first symptoms of starvation in our population, namely an increase in the many ailments that spring from malnutrition. These are evident in our crops and livestock, as well as in ourselves.

[1] McCarrison. (See p. 33.)

It is the first purpose of this book to present some of this evidence and to show that, when considering matters of health, it is misleading to separate man, animals, and plants. All are part and parcel of the same nutrition cycle which governs all living cells. If we attempt interference with it, other than along the general lines of co-operation, dire results follow. This fact is admirably explained and described in a book by Dr. K. E. Barlow, *The Discipline of Peace*.[1] But even his comprehensive survey has omitted a vital link in the cycle. He shares the common misconception that soil can be considered as an inorganic raw material. 'The dependence of the animal on the plant', he writes, 'is in respect of energy and discipline, exactly parallel to the dependence of the factory upon the steel works. Animal food is a modified product, just as metal is . . . life first of all encounters the inorganic and elaborates certain materials from it by a process which is called metabolism. As we pass from the plant to the animal cell, life no longer encounters what is inorganic; the raw material on which it now works is metabolized material. What has happened has been that a second cycle of metabolism has been integrated with the first, and with each new cycle, the scope and perfection of the discipline practised within the living organization is increased.'

This first cycle, by which the inorganic enters the organic, is certainly one of the strands in my metaphorical cable, but it is not the only one with which plant life is concerned. Interwoven with it is one of the threads of the organic cycle. The vegetable kingdom has for too long been considered as a sort of factory concerned only with converting inorganic salts into metabolized material for the benefit of the animal kingdom, without payment. It is now becoming realized, however, that to fulfil this function it demands its price. If the inorganic world supplies much of the raw material, the world of micro-organisms must supply much of the motive power; these micro-organisms in their turn depending upon the animal and vegetable kingdoms for their sustenance.

In other words, by the time food reaches the animal, much of it is entering, not its second, but its third and even fourth cycle of metabolism, for recent research has provided evidence, as I shall later show, that part at least of the food utilized by plants is a highly complex material already metabolized by bacteria and soil fungi from the organic substances of the higher plants and animals, thus completing the full cycle of life, death, decay and rebirth.

Quite a strong case can be made out for believing that it is, at least

[1] Faber & Faber, London, 1942.

in part, because we have ignored the dependence of plant life upon the action of soil fungi, that sickness today costs Great Britain, directly or indirectly, the formidable sum of £300 million sterling every year.[1]

It is as though man's ignorant and greedy exploitation of the soil had put into reverse the wheel of health. Luckily there still remain on this globe one or two notable exceptions, where the law of return is still practised; where nature's cycle is whole and unbroken, and where perfect health is to be found, in crops, in livestock and in their human consumers. Some account of these beacons of light in the surrounding gloom will be given in Chapter 7.

It is a pity that it is not part of the training of every doctor to study these oases of health in a desert of sickness, in order to discover what lessons can be learnt from them, for the mere fact that they exist, and under widely differing external conditions, suggests that the birth-right of man is full health.

What do we mean by health? Should we be content to define it as the military 'A.1' category? Does this definition imply 'health in the full dictionary sense of the word of wholeness, namely sound physique of every organ of the body without exception, and freedom from disease? . . . We want to know what is full health, whether the tremendous part illness and ailments play in modern civilized countries is really necessary, and if not, upon what primarily does health depend. We can ourselves attain to health—or at least with our modern skill in investigation we should be able to do so.'[2]

Is there any evidence that we can? I think there is, and it is the second purpose of this book to present this evidence also. Those research workers who have provided it, whether they be doctors, botanists or soil specialists, have been brought, as a result of their investigations, to an acceptance of the wholeness of the cycle of life. From whatever angle these investigators may have approached the subject, and whatever disagreements they may have had with each other on the road, they appear to have arrived at the common conclusion that a close connection exists between soil fertility and health. Here their roads separate again a little. The more cautious among them give it as their considered opinion that though there are gaps in the argument, though critical scientific proof is as yet lacking at points in it, the strength of the indications is nevertheless overwhelming.

[1] PEP Report, 'The British Health Services'.
Note: By 1956 Britain was spending over £400 million a year on the National Health Service.

[2] G. T. Wrench, *The Wheel of Health*, Daniel, Ashingdon, 1938.

Others, content to base their views on these indications backed up by circumstantial evidence, go so far as to assert their conviction that the health of man, beast, plant and soil is one indivisible whole; that the health of the soil depends on maintaining its biological balance, and that starting with a truly fertile soil, the crops grown on it, the livestock fed on those crops, and the humans fed on both, have a standard of health and a power of resisting disease and infection, from whatever cause, greatly in advance of anything ordinarily found in this country; such health as we have almost forgotten should be our natural state, so used have we become to subnormal physical fitness.

These are far-reaching claims, and it is because the implications of them are so momentous that I wish to put before you the evidence and indications on which they are based, so that, having duly considered them, you may be in a position to form your own opinion. My purpose is to present a case, but to leave the judgement to my readers.

2

Medical Evidence

[Except for the 'Medical Testament' itself and the McCarrison experiments, which are timeless, this chapter is the most dated, especially when it comes to the figures quoted for the loss of work through illness and the cost of treating the sick. The reader is asked, therefore, constantly to keep in mind while reading it, that it refers to the period 1935 to 1939. It is, however, broadly speaking, only the figures that are out of date, *not* the deductions to be drawn from them. The cost of ill health to the country today has not been reduced, but has increased, and this in absolute terms, not just because of the fall in the value of money or the increase in the population.]

It has been announced in the press that a smaller number of people are suffering from malnutrition today, under wartime conditions, than was the case in the days of peace, and that the general health of the public has not deteriorated. We are rather proud of these facts, but it is not really a matter for congratulation, for it is less a tribute to our wartime organization than a grave reflection on our pre-war standards.

It has been estimated that 31½ million weeks of work are lost each year through illness[1] and that we spend £185 million annually on curing the sick,[2] a figure representing over £3 per head of the total population of the United Kingdom. Why is this necessary?

Let us turn first to the Medical Profession for an answer to this question. The Committee representing the General Practitioners of the County of Cheshire have drawn up their answer to it in a paper which they term a 'Medical Testament'. This was published in 1939, and it shall speak for itself.

[1] *BMA Report* published by the Industrial Health Research Board. (Medical Research Council.)

[2] PEP report on British Health Services, 1936.

MEDICAL TESTAMENT

'After more than a quarter of a century of Medical Benefit under the National Health Insurance Act we, the Local Medical and Panel Committee of Cheshire, feel that we are in a position to review our experience of the system.

'Constituted by the statute to represent the panel of an area, such a Committee is in touch with all the family doctors—in the case of Cheshire some 600—within and on its borders.

'How far has the Act fulfilled the object announced in its title—"The Prevention and Cure of Sickness"?

'Of the second item we can speak with confidence. If "postponement of the event of Death" be evidence of cure, that object has been achieved: the greater expectation of life which is shewn by the figures of the Registrar General is attributable to several factors; but certainly not least to the services of the panel.

'The fall in fatality is all the more notable in view of the rise in sickness. Year by year doctors have been consulted by their patients more and more often, and the claims on the benefit funds of Societies have tended to rise.

'Of the first item, "the Prevention . . . of Sickness" it is not possible to say that the promise of the Bill has been fulfilled.

'Though to the sick man the doctor may point out the causes of his sickness, his present necessity is paramount and the moment is seldom opportune, even if not altogether too late for any essay in preventive medicine. On that first and major count the Act has done nothing.

'We feel that the fact should be faced.

'Our daily work brings us repeatedly to the same point: "this illness results from a life-time of wrong nutrition!"

'The wrong nutrition begins before life begins. "Unfit to be a mother"—from under-nutrition or nutritional anaemia—is an occasional verdict upon a maternal death. For one such fatal case there are hundreds of less severity where the frail mothers and sickly infants survive.

'The reproach of the bad teeth of English children is an old story. In 1936 out of 3,463,948 schoolchildren examined 2,425,299 needed dental treatment. Seeing that the permanent teeth develop from the 17th week of pregnancy and that certain foods, accurately known since 1918, are the condition of their proper growth, that is a reproach which should be removed. With it would go the varied host of maladies that spring from diseased teeth. That its removal is practicable is shewn by Tristan da Cunha. Most of the population of the little island, people of our race, living on the product of sea and soil, have perfect teeth which last them their lives.[1]

'Rickets, for which England was a byword when Glisson described it in 1650, is still with us. Gross deformities are rarer, but the big heads, tumid abdomens, flaccid skins, bulged joints and pinched chests are a commonplace of infancy; and even at school age, 3,457 cases of rickets with 6,415 others of spinal curvature were found in 1936 by the School Medical Officers in 1,727,031 inspections.

'Yet its prevention by right feeding is so easy that every dog breeder knows the means.

'Rickets is a heavy contributor to the C 3 population. The Maternal Mortality Committee found that there is much less in Holland where butter, milk, and cheese are plentiful and the women by virtue of their generally healthy skeletal development are protected against the risks that are commonly faced by women in the industrial areas of England.

'Nutritional anaemia is of two kinds, one subtle and apt to happen during pregnancy, the other simple and due to too little iron in the food. It is known that anaemia especially of the latter kind is common, especially among children, and women, who need much more iron in their food than men. An inquiry into the food of 1,152 families showed that 10 per cent spent 4s a week per head on food, 10 per cent spent over 14s whilst four more groups, of 20 per cent each, spent 6s, 8s, 10s and 12s respectively. The food of the three lower groups was definitely deficient in iron. It is certain from this that nutritional anaemia amongst the poorer classes is far commoner than is recognized. Here is an example: the blood colour was tested in two groups of schoolchildren, one a "routine sample" of children, the other specially selected on account of poverty. Only half the poor children and only three-quarters of the supposedly normal children had a blood colour of 70 per cent of normal.

'The final item of our indictment is constipation. Advertised aperients are a measure of its prevalence and the host of digestive disorders which result from it are a substantial proportion of the conditions for which our aid, as doctors, is sought. Yet the cause in every case—apart from rare abnormalities—is the ill choice or ill preparation of food. It is true that we are consulted on these conditions

[[1] Since greater contact with the outer world and its commercial food products, this is, alas, no longer true (1973).]

when they are established and have to deal with the effects—gall stones, appendicitis, gastric ulcer, duodenal ulcer, colitis, and diverticulitis—of years in which the body has been denied its due of *this* constituent of food or burdened with an excess of *that*. Other means of cure than proper feeding are called for at this late stage; but the primary cause none the less was wrong nutrition.

'Those four items, bad teeth, rickets, anaemia and constipation will serve as the heads of our indictment; but in truth they are only a fragment of the whole body of knowledge on food deficiencies which different investigators from Lind and Captain Cook to Hopkins and the Mellanbys have unlocked.

'But it seems to us that the master key which admits to the practical application of this knowledge as a whole has been supplied by Sir Robert McCarrison.[1]

'His experiments afford convincing proof of the effects of food and guidance in the application of the knowledge acquired.

'In describing his experiments, which were made in India, he mentions first the many different races of which the population, 350 million, is composed.

' "Each race has its own national diet. Now the most striking thing about these races is the way in which their physique differs. Some are of splendid physique, some are of poor physique, and some are of middling physique. Why is there this difference between them? There are, of course, a number of possible causes: heredity, climate, peculiar religious and other customs and endemic diseases. But in studying the matter it became evident that these were not principal causes. The principal cause appeared to be food. For instance, there were races of which different sections came under all these influences but whose food differed. Their physique differed and the only thing that could have caused it to differ appeared to be food. The question then was how to prove that the difference in physique of different Indian races was due to food. In order to answer it I carried out an experiment on white rats to see what effect the diets of these different races would have upon them when all other things necessary for their proper nutrition were provided. The reasons for using rats in experiments of this kind are that they eat anything a man eats, they are easy to keep clean, they can be used in large numbers, their cages can be put out in the sun, the round of chemical changes on which their nutrition depends is similar to that in man, and, a year in the life of a rat is equivalent to about twenty-five years in the life of a human

[1] Major-General Sir Robert McCarrison, CIE, MD, FRCP.

being. So that by using rats one gets results in a few months which it would take years to get in man. What I found in this experiment was that when young, growing rats of healthy stock were fed on diets similar to those of people whose physique was good the physique and health of the rats were good; when they were fed on the diet similar to those of people whose physique was bad the physique and health of the rats were bad; and when they were fed on diets similar to those of people whose physique was middling the physique and health of the rats were middling. . . .

' "Good or bad physique as the case might be was, therefore, due to good or bad diet, all other things being equal. Further, the best diet was one used by certain hardy, agile, vigorous and healthy races of Northern India. [Note: the Hunza, Sikh and Pathan.] It was composed of freshly ground whole wheat flour made into cakes of unleavened bread, milk, and the products of milk (butter, curds, buttermilk), pulses (peas, beans, lentils), fresh green leaf vegetables, root vegetables (potatoes, carrots), and fruit, with meat occasionally.

' "Now in my laboratory I kept a stock of several hundred rats for breeding purposes. They lived under perfect conditions; cleanliness, roomy cages, good bedding, abundant fresh water, fresh air and sunlight—all these things they had; and, they were fed on a diet similar to that of a race whose physique was very good. They were kept in stock from birth up to the age of two years—a period equivalent to the first fifty years in the life of human beings. During this period no case of illness occurred amongst them, no death from natural causes, no maternal mortality, no infantile mortality except for an occasional accidental death. In this sheltered stock good health was secured and disease prevented by the combination of six things: fresh air, pure water, cleanliness, sunlight, comfort and good food. Human beings cannot, of course, be so sheltered as these rats were, but the experiment shows how important these things are in maintaining health.

' "The next step was to find out how much of this remarkably good health, and freedom from disease, was due to the good food: food consisting of whole wheat flour cakes, butter, milk, fresh green vegetables, sprouted pulses, carrots and occasionally meat with bone to keep the teeth in order. So I cut out the milk and milk products from their diet or reduced them to a minimum, as well as reducing the consumption of fresh vegetable foods while leaving all other conditions the same. What was the result? Lung diseases, stomach diseases, bowel diseases, kidney and bladder diseases made their appearance. It was apparent, therefore, that the good health depended on the good diet more than on anything else and that the diet was only health-promoting so long as it was consumed in its entirety, so long, in fact, as it contained enough milk, butter, and fresh vegetables.

' "Many more experiments were done which showed that when rats or other animals were fed on improperly constituted diets, such as are habitually used by some human beings, they developed many of the diseases from which these human beings tend to suffer: diseases of the bony framework of the body, of the skin covering it and of the membranes lining its cavities and passages; diseases of the glands whose products control its growth, regulate its processes and enable it to reproduce itself; diseases of those highly specialized mechanisms —the gastro-intestinal tract and lungs—designed for its nourishment; diseases of the nerves. All these were produced in animals under experimental conditions by feeding them on faulty human diets. Here is an example of such an experiment: two groups of young rats, of the same age, were confined in two large cages of the same size. Everything was the same for each group except food. One group was fed on a good diet, similar to that of a Northern Indian race whose physique and health were good, and of which the composition is given above. The other was fed on a diet in common use by many people in this country; a diet consisting of white bread and margarine, tinned meat, vegetables boiled with soda, cheap tinned jam, tea, sugar and a little milk: a diet which does not contain enough milk, milk products, green leaf vegetables and whole-meal bread for proper nutrition. This is what happened. The rats fed on the good diet grew well, there was little disease amongst them and they lived happily together. Those fed on the bad diet did not grow well, many became ill and they lived unhappily together; so much so that by the sixtieth day of the experiment the stronger ones amongst them began to kill and eat the weaker, so that I had to separate them. The diseases from which they suffered were of three chief kinds: diseases of the lungs, diseases of the stomach and intestines, and diseases of the nerves; diseases from which one in every three sick persons, among the insured classes, in England and Wales, suffer."

'These researches were minutely made on a large scale and, but for the food, the conditions of each group were identical and ideal. Their results to our minds carry complete conviction—especially as those of us who have been able to profit by their lesson have been amazed at the benefit conferred upon patients who have adopted the revised dietary to which that lesson points.[1]

[1] See Chapter 6.

'It is far from the purpose of this statement to advocate a particular diet. The Eskimos, on flesh, liver, blubber and fish, the Hunza or Sikh, on wheaten chapatis, fruit, milk, sprouted legumes and a little meat; the islander of Tristan on his potatoes, seabirds' eggs, fish and cabbage, are equally healthy and free from disease.

'But there is some principle or quality in these diets which is absent from, or deficient in, the food of our people today. Our purpose is to point to this fact and to suggest the necessity of remedying the defect.

'To descry some factors common to all these diets is difficult and an attempt to do so may be misleading since knowledge of what those factors are is still far from complete; but this at least may be said, that the food is, for the most part, fresh from its source, little altered by preparation and complete; and that, in the case of those based on agriculture, the natural cycle:

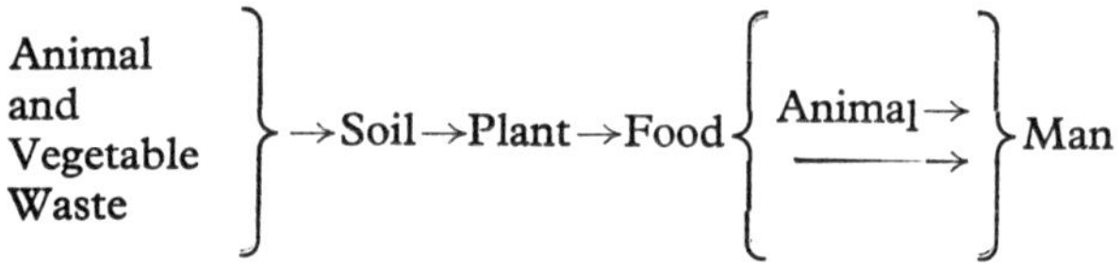

is complete.

'No chemical or substitution stage intervenes.

'Sir Albert Howard's[1] work on the nutrition of plants, initiated at Indore and carried from India to many parts of the world, seems to constitute a natural link in this cycle.

'He has shewn that the ancient Chinese method of returning to the soil, after treatment, the whole of the animal and vegetable refuse which is produced in the activities of a community results in the health and productivity of crops and of the animals and men who feed thereon. . . .

'Though we bear no direct responsibility for such problems, yet the better manuring of the home land so as to bring an ample succession of fresh food crops to the tables of our people, the arrest of the present exhaustion of the soil and the restoration and permanent maintenance of its fertility concern us very closely. For nutrition and the quality of food are the paramount factors in fitness. No health campaign can succeed unless the materials of which the bodies are built are sound. At present they are not.

Probably half our work is wasted, since our patients are so fed from the cradle, indeed before the cradle, that they are certain contributions to a C 3 nation. Even our country people share the white bread,

[1] Sir Albert Howard, CIE, MA. (See next chapter.

tinned salmon, dried milk regime. Against this the efforts of the doctor resemble those of Sisiphus.

'This is our medical testament, given to all whom it may concern—and whom does it not concern?

'We are not specialists, nor scientists, nor agriculturalists. We represent the family doctors of a great county, the county, said Michael Drayton, of "such as soundly feed"; a county which gives its name to a cheese than which there is none better, though to most Englishmen, alas, only a name; a county where the best farming is still possible, which should minister to the needs of its own industrial areas and of a far wider circle.

'We cannot do more than point to the means of health. Their production and supply is not our function. We are called upon to cure sickness. We conceive it to be our duty in the present state of knowledge to point out that much, perhaps most, of this sickness is preventable and would be prevented by the right feeding of our people. We consider this opinion so important that this document is drawn up in an endeavour to express it and to make it public.'

(*Signed by the Members of the Local Medical and Panel Committees*)

JOHN KERR (*Chairman*)
N. A. BOSWELL (*Vice-Chairman*)
J. BARRY BENNETT (*Hon. Treasurer*)
LIONEL JAS. PICTON (*Hon. Secretary*)

I think you will agree that this is an impressive document deserving of a far wider publicity than it has received. It lends point to the classic German pun 'Mann ist was man isst': 'Man is what man eats'.

Let me repeat the opening sentences of the major indictment of this remarkable testimony: 'The fall in fatality is all the more notable in view of the rise in sickness. Year by year doctors have been consulted by their patients more and more often, and the claims on the benefit funds of Societies have tended to rise. . . . On that first and major count (prevention) the Act has done nothing.'

This suggests that our national health is deteriorating; yet other authorities have published statistics to show that it is on the contrary very much better than it used to be. How can these two views be reconciled?

The following medical opinion, which is widely held, contains the answer, I think: 'the infective diseases are under far better control, but degenerative diseases—such as blood pressure diseases, heart diseases, rheumatism, diabetes, nervous disease, kidney disease, gastro-intestinal disease, cancer and mental disease—which impair the prime of life, or what should be the prime of life, are admittedly more prevalent.'

Lord Portsmouth, in his book *Famine in England*,[1] published in 1938, wrote, 'Dr. Howard Mummery, Chairman of the Association of Industrial Medical Officers, revealed the other day that workers insured under the National Health Insurance Act lost on an average twenty-eight days of work a year through illness, compared with 16·5 days of work fifteen years ago.'[2] This would seem to support the view that ill health is increasing.

Some part of this increase may indeed be due to the adoption of a different standard of 'disablement', and a greater readiness to consult a doctor. But this in itself points to a significant preoccupation with bodily states, a diminished power of resisting and throwing off incipient ailments. Certainly it does not suggest any advance in the prevention of sickness commensurate with advances in medical knowledge, Public Health services, housing and sanitation. That this view is shared by members of the medical profession in the United States of America, where health services are unsurpassed, is shown by the two following quotations:

'Great gains in health have been achieved since the beginning of this century. Tuberculosis is being vanquished. Deaths from infantile diarrhoea, diphtheria, typhoid fever, etc., are being eliminated. All diseases of bacterial origin have decreased in a striking manner. The average length of life—that is, the expectation of life at birth—was only forty-nine in 1900. Today it has gained more than eleven years.

'The chances of survival for each age up to maturity have notably augmented. Nevertheless, in spite of the triumphs of medical science, the problem of disease is very far from solved.

'Modern man is delicate. Eleven hundred thousand persons have to attend the medical needs of 120,000,000 other persons.

'Every year, among this population of the United States, there are about 100,000,000 illnesses, serious or slight. In the hospitals 700,000 beds are occupied every day in the year. The care of these patients requires the efforts of 145,000 doctors, 280,000 nurses or student

[1] Witherby, London.

[2] G. C. Anderson, Secretary of the British Medical Association, writing to the *Daily Telegraph* on 5 February 1943 stated that over 50 per cent of registered panel patients attend, or are attended by their doctors in the course of a year.

nurses, 60,000 dentists, and 150,000 pharmacists. It also necessitates 7,000 hospitals, 8,000 clinics, and 60,000 pharmacies.

'The public spends annually £143,000,000 in medicines. Medical care, under all its forms, costs about £700,000,000 yearly.

'Obviously, disease is still a heavy economic burden. Its importance in modern life in incalculable.'[1]

'Although it is true that our life span has been increased, yet an analysis of that gain shows it is confined almost entirely to lower age groups. Excluding infant mortality and the diseases of childhood, and excepting also our control of the major pestilences the chances of dying after you reach thirty-five are about as great today as they were a hundred years ago. The human race still digs its own grave with its teeth. The man who said this probably uttered one of the greatest truths. Until recently we did not know that disease could be absolutely proved to be intimately connected with diet. . . .

'Working hard when you are young and earning a competence on which you can retire and enjoy yourself sometime after you are fifty will bring you little happiness if you are not well enough to enjoy those years. Most of the diseases which we associate with this period are the direct results of an improper diet in youth.'[2]

Bearing this last point in mind, let us now examine in rather more detail the evidence of the Rats of Coonoor. Dr. Wrench in his book *The Wheel of Health*[3] discusses the results of McCarrison's experiments at some length. His observations seem to me important, and at the risk of some repetition I must now record part of what he has to say:

'In 1921 Sir Gowland Hopkins had made public his work on accessory food factors, to which Casimir Funk a year later gave the name of vitamins. McCarrison, reading the work, at once thought that maybe a very important clue to the enigma of goitre lay in a deficiency of vitamins in the food which goitrous people eat. So he began experiments in the Kasauli laboratory designed to give pigeons goitre. He fed them on diets defective in vitamins. Something different happened. The birds did not develop goitre but some of them, as was expected, developed a disease called polyneuritis. Then it was found that these birds were overrun by specific microbes. Now came the surprise. Some of the healthy birds, the stock of the laboratory who were well fed before any experiments were tried upon them also harboured these microbes, but they were not ill. The ill-fed

[1] Dr. Alexis Carrel, *Man the Unknown*, Hamilton, London, 1935, p. 114.
[2] Victor Heiser, MD, *You're the Doctor*, Cape, London, 1939.
[3] See footnote to p. 28.

birds, on the other hand, were mortally sick. If, however, the healthy birds were fed on the food defective in vitamins, they too got the polyneuritis and died. Good feeding, it seemed, protected the birds against the microbes, but faulty feeding led to a microbic triumph. Thus was McCarrison brought into a field of "deficiency diseases", that is to say, diseases due definitely to faulty food.'

The rat experiments described in the 'Medical Testament' were undertaken (after McCarrison's appointment in 1927 as Director of Nutrition Research in India) at his laboratory and headquarters at Coonoor. In the first experiment, in which he transferred the exceptional health of the Hunza, Sikh, and Pathan to his rats, 1,189 were fed on the diets of these healthy peoples, and watched from birth to the twenty-seventh month, the age which, you will remember, corresponds to that of about fifty-five years in the life of man.

Describing the results, in a lecture given at the College of Surgeons in 1931, McCarrison told how throughout the course of this two and a quarter years he had been unable to discover any case of illness in this 'universe' of albino rats, either clinically, or at post-mortem examination. Wrench commenting on this result writes:

'Now the reader might think that a statement that any small "universe" had been freed from disease would have created a profound impression amongst medical men. It did not do so, any more than Lister's announcement of the first results of antiseptic surgery created any stir. In Lister's days surgeons were so accustomed to pus and blood poisoning, they could not think in terms of surgery without them. Similarly, medical men are so accustomed to a great number of diseases, they cannot think of any small "universe" without disease. In all revolutions this is the case. It is the established profession, or class, or aristocracy, which finds it most difficult to think in terms of the change.

'This is very noticeable in the professional comments of that time upon McCarrison's lectures. Actually they were very meagre. . . . The *British Medical Journal* . . . did, however, devote a leading article. This article treated McCarrison's work purely from the point of view of diseases which diet would prevent or help to prevent. It overlooked the astonishing relation of a remarkable health of human groups being transferred to rats as a perfect health. . . . The health was transferred by foods. . . . The only thing . . . that was common to rat and man in this first experiment was the diet. Here in the great cleft of Hunza was a little oasis of a few thousand beings of almost perfect health,[1] and here

[1] See Chapter 7.

in the cages of Coonoor was a little oasis of a thousand and more albino rats also in perfect health. The only link connection between these two otherwise dissimilar sets of living things was a similar kind of diet.'

In subsequent experiments, as recounted in the 'Medical Testament', the rats were linked to other Indian peoples by their diet. In every case the average standard of health of a given human group was faithfully mirrored in the rats, including the percentage incidence of specific diseases. Nor was this reflection confined to bodily ailments, for neurasthenia, and bad-temper, you will remember, showed themselves in the rats fed on the common English diet. During the course of this series of experiments, McCarrison found, and listed, diseases of every organ of the body among the 2,243 rats fed on faulty Indian diets.[1] 'Considering . . . the simplicity of the rat, and its limitation in things human,' writes Wrench, 'the list is, comparatively speaking, almost as complete as the list of contents of a stately textbook of medicine. "All these conditions," said McCarrison, "these states of ill-health had a common causation; faulty nutrition with or without infection." '

Sir John Orr, writing about these experiments[2] points out that 'such experiments with rats, of course, do not carry the same weight as observations on human beings'. Wrench, discussing this point, writes:

'This criticism is particularly interesting because it follows, not only a brief précis of McCarrison's work, but also one of a valuable, similar and later experiment of Orr's upon rats fed upon "the average diet eaten by a working-class community in Scotland (with its) daily variation, thus mimicking the food habits of human beings". There was, however, a small addition to the quantities of milk, as rats could not be bred without the larger allowance (*Journal of Hygiene*, xxxv). The results were in the main similar to those of McCarrison.

'Now, the noticeable thing about this criticism is that Orr fragments the experiments of both McCarrison and himself. He separates the experiments upon rats from the observations on human beings.

'In actual fact McCarrison's experiments were preceded by, and due to, observations upon human beings. The men were observed first and then the rats. Orr's experiments were due to the example of McCarrison and observations of the unsatisfactory state of the Scottish working class. Ill health was transferred to rats by men's faulty food. Without the observations on man the experiments would never have been undertaken.

[1] For complete list see Tech. Ref., p. 51.

[2] *Food, Health and Income*, 1936.

'The criticism shows the hold fragmentation has upon the mental habits of scientists. Orr no sooner reaches McCarrison's truth by his own experiment than he separates himself from it owing to this habit.

'A further comment might be made, namely, that the food would never have had such a full effect if the healthy rats had not been cleanly and airily housed and enjoyed sheltered lives, even though these conditions were also those of the sickly rats.

'Air, or rather oxygen, it can rightly be maintained is a part of the food. When the human being is in the womb, oxygen is not separated in any way from the other elements of food, but is brought with them by the mother's blood. It remains a food after birth, but it has peculiar importance in assisting at so many vital processes of movement and energy that it has constantly to be sprayed into the blood by the special apparatus of the lungs.

'Hence the airy cages of the rats of Coonoor were a healthy asset, but that is all. They did not save the sickly rats. . . .

'The inescapable conclusion is that in a very large number of diseases faulty food is the primary cause. The suspicion is that faulty food is the primary cause of such an overwhelming mass of disease that it may prove to be simply *the primary cause of disease*.

'Up to the present day, it seems, the medical profession and the public have had to be satisfied with a fragmentation of causation, that is to say, a very great number of secondary causes and often enough no real causes at all, but causes as fictitious as they are popular.

'For the purposes of illustrating and emphasizing the really immeasurable importance of this contrast, if correct, let us take some few of these illnesses and put their causes as given in medical textbooks and as shown by the rats of Coonoor in parallel columns.

'Let us first take that dangerous disease, pneumonia. Pneumonia is due to a microbe, the pneumococcus, which is found in masses in the lung in true lobar pneumonia. The pneumococcus, says the textbook, is a resident of the human mouth. It is found in 80 to 90 per cent of normal healthy individuals. Something more, then, than the mere presence of the pneumococcus must be the cause of pneumonia; something that makes this domesticated microbe suddenly become dangerous. In other words, the pneumococcus cannot be called the primal cause of pneumonia. Something has to precede it—some weakness of the barrier.

'The weakness of old age is given first of the orthodox causes in the textbook. . . .

'Pneumonia is more common in cities than in the country and in males than in females. Any weakening habit, such as that of over-drinking, becomes a cause, and also makes the microbes more lethal. Yet robust men may be attacked. Cold is a cause if it weakens, but not if a man finds it a tonic and reacts to it. A previous attack makes a second attack more probable. Another illness, such as chronic kidney or heart disease or one of the acute infectious fevers, gives opportunity to the pneumococcus. So also pneumonia may follow a blow on the chest.

'Now let us place these causes from textbooks in juxtaposition with that of the "small universe" of Coonoor.

CAUSES OF PNEUMONIA

Textbooks	*Coonoor*
Weakness of old age.	Faulty food.
Debilitating habits.	
Exhaustion.	
Chill.	
Previous attack.	
Some other illness, chronic or acute.	
A blow on the chest.	
Pneumococcus microbe.	

'. . . Summing up the textbooks' causes, one may call them a number of added weaknesses to an inferior barrier against disease. The barrier gives way readily at this or that point. In other words the barrier has degenerated.

'By his skilled science man is actually able to get a partial picture of what this barrier is. It is in fact an actual barrier. It can be seen through the microscope. It can be seen if it looks healthy or degenerate. It can be photographed, and the photograph of a healthy barrier has clear outlines and demarcations and that of a degenerate barrier is blurry. This barrier is the fine skin which lines the tubes and cells of the nose, windpipe and bronchial tubes, of the mouth, throat, stomach and small and large gut. The fine interior skin is much the same as the outer skin of the body, only it is thinner and softer. But both have an outer layer of cells called epithelium, and it is the epithelium that can be particularly well seen under the microscope. It is the epithelium that forms the visible barrier and which shuts out microbes and other intruders. It does not by any means form the whole barrier, but it constitutes a part of it, which can be seen as clear and definite or blurred and indefinite, according to whether it is itself well or ill fed. The contrast picture gives anyone with even a little knowledge of the microscope a good idea of what can be termed the barrier or, more accurately, the first line of defence. It is not fiction.

'So we can understand how it is that faulty food can stand alone under the heading Coonoor against the juxtaposed textbooks' list of the causes of pneumonia. It can be placed there as primary, and thereby able to make all the causes in the textbooks possible; it can activate them. Without it they would be inert.

'Now let us look at the common infection of the middle ear. Mellanby found this infection in a fifth of his faultily fed rats. It was common among the ill-fed rats of Coonoor, but absent in the well-fed. On the other hand, a well-known textbook, such as Politzer's *Diseases of the Ear*, does not mention faulty food as a cause, any more than faulty food was mentioned under pneumonia. That the whole basis of modern life may be wrong and that that is why such large textbooks have to be written has not as yet appeared in the textbooks themselves.

'Putting the causes of acute infection of the middle ear into juxtaposed columns, we have:

Textbooks	*Coonoor*
External atmospheric conditions.	Faulty food.
Colds in the head.	
Infectious diseases, such as measles, pneumonia and influenza.	
Sea baths.	
Nasal douches.	

'There must be, therefore, in faultily fed people, a fear of cold night air, colds in the head; other people coughing and sneezing; schools where children mingle with children; bathing in the sea; and keeping off the "flu" by snuffing lotions or using nasal douches, as recommended by advertisers. Any of these things may lead to passing of the barrier and the defences of the tissues of the ear.

'The eyes are even more commonly affected by faulty food than the ears. The sickly rats of Coonoor got inflammations of the eyes, ulcers, and a particular "dry" eye leading to blindness. All of these the well-fed rats escaped.

'The textbooks all accept defective food as a cause of "dry" eye, of xerophthalmia, and recommend cod liver oil and butter, which will cure it if not too far advanced. With this exception there is no direct reference to faulty food as a cause of diseases of the eye. There is only the general statement that these diseases are more common among the poor and debilitated.

'A medico-surgical disease which is of particular interest is peptic ulcer, or ulcer of the stomach or duodenum. It is of particular interest because of its proven direct relation to faulty food. It happens to be very common amongst the poorer classes of Southern Travancore—so common that both Lt.-Col. Bradfield, IMS, and Dr. Somervell asked McCarrison to put rats on the foods as prepared and eaten by these people.[1] He put a batch of rats on the food as prepared and cooked by the poorer folk of Southern Travancore for 675 days, and at the end of that time peptic ulcer was found in over a quarter of them.

'This striking result has not yet appeared in the textbooks. As is the way of new knowledge, it passes into currency by a process of slow percolation. Until the time comes when it reaches the textbooks the causes of peptic ulcer placed in juxtaposition, appear as follows:

CAUSES OF PEPTIC ULCER

Textbooks	*Coonoor*
Occupation: anaemic and dyspeptic servant girls, shoemakers, surgeons.	Primarily faulty food.
Injury.	Specifically such food as that of the poorer classes of Southern Travancore.
Associated disease such as anaemia, heart disease, disease of liver, appendix, gall bladder, teeth, tonsils.	
Nervous strain.	
Disturbances of the circulation.	
Large superficial burns.	
Certain families are said to be more liable.	
Increased acidity of the stomach.	
Several of the above in combination.	

'The last disease I propose to take in these few illustrations is tuberculosis. As regards this dreaded disease, McCarrison, in the Cantor Lectures, turned from his own work to one of the most remarkable of human experiments, that of the Papworth Settlement, so intimately associated with the name of Sir Pendrill Varrier-Jones.

'Papworth is a settlement for sufferers from tuberculosis, mostly in the form of consumption of the lungs. The patients are, of course, ill when they come to the settlement, but under a care, really quite like that given to the rats of Coonoor, namely adequate food supply, good housing, and ventilation and freedom from anxiety in the form of loss of employment, there are remarkable and sustained recoveries.

'All patients at Papworth have sputum pots and pocket flasks into

[1] This contains a high proportion of tapioca.

which they must spit. The infected sputum is at once made innocuous. Moreover, public opinion in the village enforces their use by attaching shame, not to the users, but to those who dare to be forgetful.

'In Papworth there are many married couples. The children of these couples live in the settlement. They are in frequent contact with tuberculosis and are protected from the disease by the general use of the spitting pots and flasks and by good food, or in Varrier-Jones's own words: "the child's resistance to disease is maintained by

(a) adequate nutrition, and
(b) the absence of mass dose of infection."

'Now comes the outstanding fact. The Papworth village has been in existence twenty years, yet not one of the children of these married couples has developed any form of tuberculosis. "Our experience proves," writes Sir Pendrill in his report for 1936, "that no tuberculosis disease need be transmitted so long as village settlement conditions of housing and employment are properly utilized. Any question of 'heredity' is now generally discredited."

'In face of this testimony to the power of resistance to tuberculosis given by good food and housing, and with spitting pots to avoid mass infection, the textbooks put forward "pre-disposition" as a widely accepted medical tenet. . . .

'This terrible Calvinistic doctrine, by which certain people, and particularly artisans of the cities, are born predestined to get tuberculosis has therefore been challenged by the good food, security, and the avoidance of mass infection at Papworth.

'The Papworth results suggest the following juxtaposition:

CAUSES OF TUBERCULOSIS

Textbooks	*Papworth*
Infection with tubercle bacilli.	Inadequate nutrition.
Inherited predisposition.	Mass dose of infection.
Living in dark, close alleys, and tenement houses, excess of alcohol and other weakening habits.	
Confinement in prisons, workhouses and workshops.	
Catarrh of respiratory passages.	
Diabetes, kidney disease and other chronic affections which lower resistance.	

'If the list of the textbook is carefully examined we see how the causes there given are all, except that of diathesis, to be found contained in the two Papworth causes. Infection with the tubercle bacilli in the one column is duplicated by the mass infection of the other. Frequent inhalation of quantities of the microbes gives greater opportunities to them to breach the barriers. All the rest are the fragmentation of "inadequate nutrition".

'Living in dark, close alleys and tenements means also faulty food. The impure air of slums means one food, namely, oxygen, being defective, but it means also that people who breathe it have not the money for foods that cannot, like oxygen, be got for nothing. Alcohol in excess destroys the appetite. So do the poisons of such diseases as diabetes and kidney disease. So does confinement in prisons, workhouses, and workshops. None of the people debilitated by such places or such diseases eat heartily of good food. As to catarrh of the respiratory passages, that in itself was produced by McCarrison and also by Mellanby by faulty food. The barrier breaks down before the catarrhal microbes. A mass attack of tubercle bacilli may do the rest.

'If then, one can put aside the predestination theory of tuberculosis, there lies one thing behind all the other causes given, and that is faulty food and, moreover, as we shall see, faulty food may account for the apparent predestination.

'Fortunately there is another triumph in establishing the general cause of many diseases and ill health in poor English children. With just as unpromising human material as that of the Papworth children the late Miss Margaret McMillan gained this success, which is described in her book *The Nursery School* (1930).'

This is an open-air school in Deptford, and the children are drawn from the surrounding slums. It is on a sufficiently large scale for the results to have critical importance. During the years described in Miss McMillan's book, it consisted of 260 children from the ages of two to five, of which eighty continued under her care, until they reached the age of fourteen or fifteen. Dr. Wrench does not minimize the importance of the all-round care devoted to these children, but obviously regards 'well considered food' as being one of the most important factors in the truly remarkable results achieved. 'Next door to the school,' he writes, 'is "our own" Deptford Clinic for sick children. School and clinic under that one authority present themselves as human replicas of the rats of Coonoor.

'Here is Miss McMillan's description of the food of the school. "Out all day in moving air, children are always hungry at mealtimes, but no food is given between meals. In summer they have fruit from the old mulberry tree, and we give small spoonfuls of orange juice. Fruit and fresh vegetables are needed by everyone, but especially by growing children, and most of all by children of the poorest classes in cities. Their bones are literally starved of mineral salts. They suffer from starvation in the way of nitrogenous food and of all that nature supplies in green food and fruits. Bread, bread, and always bread in surfeit is their portion. Our fresh vegetables, meal and milk work wonders."

'The test of a diet is the wholeness of those who eat it. This is the description of the children of seven, after four years spent in the school: "They are all straight, well-grown children, alert, sociable, eager for life and new experience. . . . The abyss between him and the child of yesterday yawns deepest, perhaps when we compare the *state* rather than the achievements of the nurtured child with that of the other. The nurtured seven-year-old is a stranger to clinics; he knows little about doctors. He sees the dentist, but hardly ever, or perhaps never, needed any dental treatment."

'To "our clinic" come the sick children of Deptford. They are just ordinary poor children who go to other schools and have other homes than hers. They present the picture of the sickly rats of Coonoor; Miss McMillan draws the contrast, though not in juxtaposed columns.

' "There, ranged on seats by the walls, sit scores of sufferers. Blepharitis, impetigo, conjunctivitis, skin diseases of many kinds, these are not seen in our school. They are seen in the clinic—thousands of cases all preventable." There follow further illnesses seen in the clinic—adenoids, bad tonsils, colds, coughs, bronchitis, enlarged glands, gastric and intestinal troubles—in short, the list which afflicted the sickly rats of Coonoor.

'Now both Sir Pendrill Varrier-Jones and Miss McMillan have been exceptionally imaginative in seeing that *all* the conditions of life in those under their care were made wholesome, things of the mind as well as those of the body, and it is to this wholeness that they attribute the health of their wards. They do not select food as the primary cause of the health. They regard the whole as resulting in health.

'This is so reasonable that I think no one reading their results would care to diminish any one guard of healthy life which they have erected, such as modern housing and hygiene.

'Yet, apart from proofs and arguments already put forward to maintain the vital primary claim of food, there is one very exquisite human experiment made by Dr. G. C. M. M'Gonigle, Medical Officer of

Health of Stockton-on-Tees, which strengthens this claim in a manner that may be called one of accidental finality.

'Stockton-on-Tees is an ancient market town which has grown rapidly in the last three-quarters of a century and now has a population of 67,722 (1931). Of this population in that year 40 per cent of the males between fourteen and sixty-five were unemployed.

'Stockton has slums, and the Town Council recently carried out a vigorous policy of better housing. It was this that gave M'Gonigle an opportunity to exercise his excellent powers of scientific observation.

'A survey of housing needs was taken in 1919, and the largest section of the town scheduled as an unhealthy area was dubbed "Number 1 area". It was decided to demolish a part of Number 1 and transfer its inhabitants to a new up-to-date municipal estate, agreeably named Mount Pleasant. In 1927, 153 families, comprising 710 individuals, were transferred to Mount Pleasant, leaving behind in Number 1 area 289 families with a total of 1,298 individuals.

'Here, then, were contrasting conditions of new and old, of good housing and slum. Naturally everyone thought the transfer to Mount Pleasant would be a betterment. But M'Gonigle watched.

'Even he, however, watched at first according to the routine of his official position. It was only when he found that something odd was happening and the expected success was not coming off, that he concentrated a keen and skilled observation upon the anomaly.

'His attention was drawn to it by the fact that the health of the inhabitants of Mount Pleasant, instead of improving or at least remaining stationary, began to deteriorate, whereas that of those families and people left behind in the slums did not.

'M'Gonigle then began to test out what was happening statistically. The standardized death-rate of the first five years following upon the transfer was 33 per 1,000; that of the unchanged slum 22 per thousand. The rate for the Mount Pleasant estate of "33·55 per thousand, appears to be extraordinary, in view of the fact that it represents an increase of 46 per cent over the mean standardized rates for the same individuals in the previous quinquennium," is M'Gonigle's comment. The increase was not due to any peculiarity of infant mortality, epidemic, or other recognized cause. It was just there steadily throughout, and it represented an increase in the various groups, from 0 to 10, between 10 and 65, and over 65. It was a characteristic of the whole people of Mount Pleasant. It was "a real increase and beyond the probable extent of fortuitous variation".

'What was it due to? The better housing? It seems absurd that something better should prove something worse. Yet, in spite of the best intentions, this happens if primary things are forgotten. Man lives primarily by food, not by housing, and the food of the Mount Pleasant people was what had deteriorated.

'When living in the slums these people paid rents which averaged 4s 8d a week per family. In 1928, on the Mount estate, the rent was 9s a week, and by 1932 it had risen to 9s 3½d per week, or double the original rent.

'Consequently there was less money to spend on food.

'M'Gonigle worked out the average amount spent on food per individual for Mount Pleasant and for slum by carefully prepared and corrected statistics. It is obvious, in view of the different rents paid, that Mount Pleasant was worse off. Particularly was this shown in the case of unemployed of both areas. The food per "man" per week at Mount Pleasant cost 34·7 pence, that in the unchanged slum 45·6 pence.

'M'Gonigle was, therefore, forced to the conclusion that the deterioration of food led to the deterioration of health. "Such environmental factors as housing, drainage, overcrowding or insanitary conditions" could obviously be excluded. These secondary factors were not worse at Mount Pleasant. They were a great deal better. That was the good fortune of this illuminating experiment. The secondary things, namely, housing and sanitation, were made better first, and in making them better money was withdrawn from the individual's primary need—food.

'The experiment emerges as an example of putting the building of new houses and of organizing physical drill on a par or as prior to food in policy of health. They are both good things, but they are not primary.

'Muscular energy and activity follow right feeding naturally, and physical training can follow upon the muscular energy. No one indeed disputes this proposition—except in their acts and public policies. There is a general, rather indefinite feeling that sound food is the primary cause of health, but when this shapes itself out of the mist, there appear secondary, not primary forms—good housing, hygiene, physical drill.

'M'Gonigle showed that food took the primary place to good housing and sanitation.'

I would ask you specially to note this last point in view of the number of prominent people who today give housing *first* place in their plans for reconstruction.

Technical Reference

DISEASES FOUND IN RATS OF COONOOR AS A RESULT OF FEEDING THEM ON FAULTY HUMAN DIETS

Lung Diseases: pneumonia, broncho-pneumonia, bronchiectasis, pyothorax, pleurisy, haemothorax.

Diseases of the ear: otiis media or pus in the middle ear.

Diseases of the nose and accessory sinuses: sinusitis.

Diseases of the upper respiratory passages: adenoid growth.

Diseases of the eye: conjunctivitis, corneal ulceration, keratomalacia, panophthalmitis.

Gastro-intestinal diseases: dilated stomach, gastric ulcer, epithelial new growths in the stomach, cancer in the stomach (in two cases only), duodenitis, enteritis, gastro-intestinal dystrophy, statis.

Diseases of the urinary tract: pyonephrosis, hydronephrosis, pyelitis, renal calculus, ureteral calculus, dilated ureters, vesical calculus, cystitis, incrusted cystitis.

Diseases of the reproductive system: inflammation of the uterus, ovaritis, death of the foetus in utero, premature birth, uterine haemorrhage, hydrops testis.

Diseases of the skin: loss of hair, dermatitis, abscesses, gangrene of the tail, gangrene of the feet, subcutaneous oedema.

Diseases of the blood: anaemia, 'pernicious type of anaemia', Bartonella Muris anaemia.

Diseases of the lymph and other glands: cysts in the submaxillary glands and accessory glands in the base of the tongue, abscesses in the same, and occasionally also in the inguinal glands, enlarged adrenal glands, atrophy of the thymus, enlarged mesenteric, bronchial and other lymph glands.

Diseases of the nervous system: polyneuritis.

Diseases of the endocrine system: lymph-adenoid goitre, and very occasionally, haemorrhage into the pancreas.

Diseases of the heart: cardiac atrophy, occasionally cardiac hypertrophy, myocarditis, pericarditis, and hydro-pericardium. 'Oedema'.

3
Humus

If 'faulty food' is the correct reply to our question: 'Why so much sickness?'—if we are, in fact, what we eat—then the logical next question is: 'What is the matter with our food?' For while it seems clear that much of the blame must rest on our often ill-chosen diet, the complete answer must go deeper than this. In the USA, for example, a very great deal of attention is paid to the proper balancing of diet, and milk, green leaf vegetables, and raw salads (McCarrison's 'health foods') are eaten in large quantities, yet America's '100 million illnesses a year' does not suggest that her people are markedly healthier than we are on that account. Dr. Alexis Carrel states, 'Modern man is delicate'. If the primary cause of this decline in vigour is faulty nutrition, then the natural conclusion is that there must be something lacking in the quality of our foods themselves; something which was not lacking in the foods of our more robust forefathers.

In this chapter I shall present the evidence supporting the view that this conclusion is correct. Some part of this deterioration in food is, indeed, readily demonstrable and is due to the commercial development of 'processed food' wherever this has involved submitting it to treatment which profoundly alters its biochemical constitution.

The case of white flour is an outstanding example of such alteration. In removing the germ, modern methods of milling have removed those parts of the wheat berry which contain the vitamins B and E, a removal which, until the introduction of the National loaf and except for a brief period during the last war, has been complete since 1872. The everyday results of this are well illustrated in the following anecdote from Adrian Bell's *Men and the Fields*.

'I found a man of over seventy cutting up a fallen tree. . . . He used a curious phrase to justify the conditions under which the men of his father's generation had worked. "Well, they had their life," he said. "Mind you, I couldn't work today like they worked—not if I was young, I couldn't work like I worked as a young man. The bread today hasn't got the stay in it. I know, because I've worked on it. When I used to go to work and we baked at home, when I'd had my breakfast that'd stay by me to dinner-time. But when we took to bakers' bread, why, after you'd worked for an hour that'd be gone and you'd feel faint inside." '

Expressed in more scientific language:

The introduction of steel-roller mills nearly seventy years ago has 'resulted in the reduction of the nutritive value of the protein, in serious lowering of the content of calcium, phosphorus and iron, in reduction of the vitamin B_1 and vitamin B_2 complex content, and probably in complete removal of the vitamin E, all representing dead loss nutritionally.'[1]

When it is remembered that B_1 is the vitamin essential to the proper nourishment of nervous tissue, and that vitamin E plays an important part in reproduction and probably also in general vigour, there is cause for reflection upon the increase of nervous disorders and the falling birth-rate in the present generation. May not these conditions, too, be attributable to faulty diet?

It is generally admitted that wheat germ oil is one of the richest known sources of fertility. It is widely used in both human and animal medicine in the treatment of sterility.[2] Its restoration to our national loaf may well benefit the nation in more ways than in the mere saving of shipping space.[3]

Further possible evils resulting from food processing are suggested by Dr. F. J. Poynton, who, writing in the *British Medical Journal* of 21 October 1933, records his opinion that infantile scurvy is on the increase. He reports no less than three cases at one time in his ward at the Great Ormond Street Hospital for children.

'This seems remarkable,' he writes, 'when we realize the great work that has been done on vitamins, and when we look upon the position of vitamin C as one of the best understood among them.' Discussing the possible causes of present-day infantile scurvy, he writes: 'I am not prepared to do more than direct attention to the vast quantities

[1] A. M. Copping, 'The Nutritive Value of Wheaten Flour and Bread', *Nutrition Abstract and Reviews*, 1939.

[2] 'In 1932 Vogt Muller treated 20 sterile women with an extract of wheat germ oil, and 17 of them had living children. Dr. Currie, of Leeds, in 1936 related success in curing habitual abortion, while in 1940 C. G. Collins (et al.) cured 21 out of 25 similar patients. In both series of cases wheat germ oil was the main factor in the cure.' Quoted from letter to *The Times*, 1 May 1942, by L. J. Picton, OBE, BA, BM.

[3] The rise in the birth-rate for 1942 is significant in this connection.

of dried milk now in use and to raise a question which I have raised before, whether such foods even given with precautions, make for the best constitutions in years to come. We know that some children do not take to fruit juices well, or are thought not to take them well, and these are then discarded; and should infantile scurvy really be on the increase, and my experience not be only a hospital coincidence, it is clear to me that many diets must touch the border line of a pathological metabolism.'

'Note. The insecurity of reliance upon orange juice to correct the damage done to milk by heating it (pasteurizing, boiling or drying) is rendered more obvious by the common use of cold storage for oranges, as, according to Professor Plimmer, vitamin C is slowly destroyed by freezing.'[1]

'Dr. Evelyn Sprawson, London Hospital, stated that children in the institution in which he worked who were fed on raw milk had perfect teeth, whereas others in circumstances identical in all respects except that their milk was pasteurized, had defective teeth.'[1]

This is a point worth underlining in view of the present agitation in favour of compulsory pasteurization.[2] There would appear to be some, as yet undefined, harm done to foods by long storage of any kind, whether frozen, heated and tinned, or dried (probably least by drying).

McCarrison states:

'There is something in *freshness* and *quality* of foods which is not accounted for by the *known chemical* ingredients of food: proteins, fats, carbohydrates, minerals and vitamins.'[1]

'It is certain that this "something" plays a part in that perfect combination of eye, muscle, nerve, blood-vessels and endocrines, which enables the heron to avoid the hawk; and in that other protective equipment—serenity and courage, clean blood plasma, and lively reticulo-endothelial system—which wards off infection and constitutes a natural active immunity.'[3]

Systems of processing and storage, however, are not the only factors responsible for deterioration in the vital quality of human food, for these foods themselves, when in the living state, show a decline in resistance to ailments; a lowered 'quality' and vigour exactly parallel to that found in the human race.

[1] Quoted from the [original] References to the 'Medical Testament'.

[2] [For a brief discussion of the pros and cons of this controversial topic see my note at the end of this chapter.]

[3] L. J. Picton, 'The Problem of Natural Food', *Medical Press and Circular*, 8 November 1939.

The loss to the livestock industry caused by such complaints as mastitis, contagious abortion, Johne's disease, foot-and-mouth disease, swine fever, swine erysipelas and white scour—to mention only a few—runs into millions of pounds annually and is on the increase. I well remember, for example, that when I was an agricultural student in 1915 we were told by our lecturer on animal hygiene that foot-and-mouth disease need not concern us, since the disease was obsolete in this country. It is commonly held that the prevalence of foot-and-mouth in the last decade is due to imported meat. Col. G. P. Pollitt, in his little book *Britain Can Feed Herself*,[1] expresses this view. 'The repeated outbreaks in this country,' he states, 'are primarily, if not wholly, due to our imports of uncooked meat.' I find this explanation unconvincing; the imports of meat into this country in 1913, shortly before the disease was said to be obsolete, amounted to 1,163,911 tons.[2] In 1936, when it was most emphatically not obsolete, imports were 1,404,069 tons,[3] only 240,158 tons more, while in the summer of 1942, when imports must have been very much less, there were nearly 700 outbreaks of foot-and-mouth.[4] When the prevalence of this disease is considered in conjunction with the increase in most other diseases, it is impossible to avoid the conclusion that our livestock is less robust than it used to be.

Disease in crops shows a similar upward grade. Potato diseases of various kinds, 'Take-all' in wheat, blight, white fly and other parasites, and innumerable fungus and virus diseases are increasingly rife.

This close parallel in the decline of health in crops, livestock, and man is very significant, and evidence will presently be given indicating that the principal cause in all three is the same—faulty nutrition.

The food of our crops and domestic animals has undergone as marked a change during the last hundred years as that of man, and the change has been of a similar order.

'We are so made as to derive our sustenance from what was alive so recently that, somatically, it is still living', writes Dr. Picton.

It seems that it is not only desirable for our own food, whether

[1] Macmillan, London, 1942. [2] Board of Trade. [3] Whitaker, 1938.

[4] Since this was written certain information has been given to me, but which I am, unfortunately, not permitted to publish, which lends considerable support to Col. Pollitt's view that outbreaks of foot and mouth disease are due, in the first instance, to imports of uncooked meat. But even if this should be proved beyond doubt it still would not suffice to explain why farms in the middle of an infected area escape infection, while others with less direct contact succumb. Sound feeding will not of course remove *sources* of infection, but may, and probably does, give protection against anything less than mass doses of infection. Further evidence on this point is given in Chapter 6.

animal or vegetable, to be as fresh as possible, but for those foods themselves also to be nurtured on a living diet. Until the war forced us to look once more to our own soil for the sustenance of our farm animals, this all-important element of 'freshness' had to a large extent been eliminated from the feeding of our livestock by the introduction of crushed seed cakes, and other imported processed foods. And the crops which we grow, whether for direct human consumption or as food for our livestock, are likewise being fed on substitutes for their natural diet of *living* organic matter.[1] This change has resulted in our soil being seriously denuded of its humus content, with a consequent loss of fertility.

History suggests that a decline in soil fertility is always accompanied by a corresponding decline in the vigour of the people who dwell upon it.

To those who doubt this, I recommend a study of the fall of the first Roman Empire. The point was stressed quite recently in a broadcast by the Minister of Agriculture who affirmed that the basis of a strong nation lay in the fertility of its soil. The necessity today is to redefine the word fertility. I would remind you of a sentence in the Survey of Soil Erosion, quoted in the first chapter, that 'Increase of production must not be confused with increase of fertility. Increased production for human use can be, and usually is, secured by cashing in on existing fertility and using it up with the disastrous effects described.'

This is what has been happening throughout the world, but in this country the process did not really begin in a way that was serious until about 100 years ago, when chemical fertilizers were introduced. Before that time soil fertility was maintained by a combination of suitable crop rotations, bare fallows, and the application to the land of various forms of organic matter, principally farmyard manure. In 1840 the famous German chemist, Liebig, published an essay, 'Chemistry in its application to Agriculture and Physiology', which has profoundly affected Western civilization. In this he propounded his theory of mineral plant foods. Roughly this rather naïve theory is that everything required by a living plant is to be found in the mineral salts present in the ash of such a plant after all organic matter has been destroyed.

It seems a curious thing that this theory should have gained such ground as to overthrow all the experience and practice of the ages.

This can only have come about through the confusion of thought

[1] See next chapter.

with regard to fertility mentioned above. Man's understanding was blinded by the increase in production which the application of this theory at first brought about. Only the true peasant, the man who, despite all modern agricultural science, still has a truer understanding of the soil than any theorist, was not taken in. He shook his head and foretold the evils which were to come. Once more his views are well expressed in Adrian Bell's *Men and the Fields*:

'He had further ideas about food. "If people ate more of what's grown with muck, there'd not be half the illness about. People say that what's grown with artificial manure does you as much good as what's grown with muck. But I know that's wrong. What's grown with chemicals may look all right, but it ain't got the stay in it.' "

But the age of science had dawned. Intuitive judgements were 'un-scientific' and 'old-fashioned'. Liebig's theory could be 'proved'. Ocular demonstration was what counted. A given quantity of these new fertilizers produced the promised increase in yield. The methods of our forefathers were clearly out of date. 'There is a prevailing popular tendency to seek to replace "intuitive" methods, wherever they may be found, by "scientific" procedures. In this attempt we sometimes find that anything which can be dubbed "intuitive" is treated as though it were simply baseless speculation, whereas mere planned communicable routine often glories in the name of "science".'[1] This tendency may have helped to establish the new soil science, in any case it was steadily gaining ground when it received a greatly quickened impetus after World War I, through the necessity which faced those manufacturers of explosives, whose factories were equipped for the fixation of atmospheric nitrogen, to find other markets for their products. This resulted in the manufacture and use (assisted by a vast advertising campaign) of huge quantities of sulphate of ammonia and other synthetic fertilizers.

For some time, while the original supply of humus in the soil lasted, all seemed to go well, and in the difficult days of cut-throat competition that intervened between the two wars, when quantity rather than quality became the standard of efficiency, cultivators generally formed the habit of basing their manurial programmes on the cheapest forms of nitrogen, phosphates, and potash on the market.

All this time the 'man of the fields' remained unconvinced. 'Wait'; he said, with that patience only close contact with the soil can give, and now the indications are mounting that the day of his vindication is at hand, for there are signs that yields are declining, and that

[1] R. C. Oldfield, *The Psychology of the Interview*, Methuen, London, 1941.

increasing quantities of fertilizers are required to produce a given return. But this is not all; quality is declining even quicker than yields. Fertility, as will presently be shown, depends on humus. The accelerated growth induced by chemical fertilizers has the effect among others, of speeding up the rate at which humus is exhausted. As this depletion of humus proceeded, troubles began. Parasites and diseases appeared in the crops, and epidemics became rife among our livestock, so that poison sprays and sera had to be introduced to control these conditions.

A very clear example of this is to be found in the case of some of the Lincolnshire potato lands.

The siltland area south of the Wash, now famous as the chief potato-growing district in England, was, until 100 years ago, equally famous for its pasture. This was so fertile that it was capable of producing prime fat bullocks without the assistance of any additional corn or cake. When these pastures were first ploughed up they produced abundant crops of healthy potatoes of good-keeping quality without the aid of fertilizers or poison sprays. Potato growing became so profitable that whole tracts of fertile grassland were turned over to the production of arable cash crops, with the result that farming became thoroughly unbalanced. Gradually chemical fertilizers became necessary to maintain yields, then diseases made their appearance, and spraying with copper salts had to be introduced to control blight. Today it often happens that all the potato tops have to be destroyed with sulphuric acid before they are mature, in order to 'preserve' the tubers underneath. Lately a new trouble, eel worm, has become alarmingly prevalent. This, too, the experts are attempting to check by chemical means. The fallacy of this method when applied to eel worms will be shown in Chapter 6.

In recent years some Lincolnshire farmers have abandoned the use of fertilizers in favour of organic composts, with the result that health has been restored to the crops and the use of poison sprays is becoming unnecessary. I do not intend to suggest by this that all so-called artificial fertilizers are necessarily harmful in themselves, but as will be shown in the next chapter, other standards than those of chemical composition alone, are needed with which to assess their value. In any case, harmful, harmless, or beneficient, their use without adequate supplies of organic matter can never serve to replace the top soil which is mechanically removed from the land by the lifting of root crops. Such crops not only take from the soil the nutrients required to grow them, but large quantities of actual soil are carted bodily from the field in the process of harvesting. If this is not replaced by organic matter, the soil loses its 'life' and with it the humic cement which holds it together. It then becomes subject to erosion by wind or rain.

'The illusion that fertility can always be restored by applying some of the huge amounts of artificial fertilizers now available has been shattered by the recognition that fertility is not merely a matter of plant-food supply (for even exhausted soils usually contain ample reserves of plant-food), but is also closely connected with soil stability. An exhausted soil is an unstable soil; Nature has no further use for it and removes it bodily. . . .

'The earlier stage of erosion is loss of fertility. Whatever the cause of the loss, the result is invariably a corresponding loss in soil stability; the soil is deprived not only of its productive power, but also of its capacity for remaining in place. Fertility is a term that should be applied to the soil and vegetation together, for the soil derives its capacity for producing life from the vegetation, as much as plants derive their capacity for growth from the soil. Apart from the indispensable plant-food elements and humus returned to the soil by the dead vegetation, the living vegetation protects the soil in many subtle ways from the erosive effect of wind and rain. . . . The capacity to absorb and retain water is a very characteristic property of mature, fertile soil. It is scarcely, if at all, developed in bare weathered rock formations (except heavy clays) that have never carried vegetation, and contain no humus. Normally the water-holding capacity of soil is confined mainly to a few inches on the surface where fresh humus, formed from decaying plant and animal remains, accumulates. Sheet erosion, by removing the most absorbent layers, not only greatly increases the amount of run off water which is the principal eroding agent, but equally decreases the value and usefulness of the rainfall. . . .

'The processes of wind erosion are less complex than those of water erosion. The predisposing conditions for wind erosion are (a) an absence of protective covering for the soil and a low fertility level, causing the soil to pulverize; (b) a dry period, as wet soil does not blow appreciably; and (c) a broad, flat or slightly undulating region across which wind and soil can move unhindered. Nature has arranged that where water cannot punish man for his ignorance and misdeeds, the wind can. When a large open area has been consistently mismanaged and its fertility reduced below the safety point, wind erosion can produce chaos within a few days.'[1]

[1] G. V. Jacks and R. O. Whyte, *The Rape of the Earth*, Faber & Faber, London, 1939.

This was very noticeable in 1942 on the sugar-beet lands of Lincoln, Norfolk and Suffolk. This crop is nearly always grown with large quantities of fertilizers, and in many areas of these three counties, the conditions for wind erosion have been produced. During the period of high winds which occurred in May of that year, very many acres of land in these three counties under sugar-beet and carrots had their top soil blown clean away, carrying the crop with it. Some farmers had to sow as many as three times.

The appearance, in a crop, of parasites and disease is often the first symptom of that loss of fertility which, if ignored, is liable to lead to lack of soil stability. Such visitations therefore should be regarded less as an enemy to be attacked direct, than as a danger signal, warning the cultivator, while there is still time to repair the damage, that a serious misuse of the land is taking place.

'The policy of protecting crops from pests by means of sprays, powders, and so forth is unscientific and unsound, for even when successful, such procedure merely preserves the unfit and obscures the real problem—how to grow healthy crops.'[1]

We shall see that the need for such protection seldom arises in those areas of cultivation where the tradition of peasant farming has never been broken, for the keystone of that tradition is the return of all wastes to the land. In almost every other type of farming, the soil has, during the last 100 years, been losing its organic matter faster than it is being replaced. This is due, not only to the excessive quantity of chemical fertilizers now in use, although this has accelerated the process, but also to the shortage of available organic matter which modern standards of civilized society have brought about, the three main factors being, the introduction of the petrol engine, the movement of the rural population into large urban areas (which the industrial revolution induced) and, perhaps most serious of all, modern sanitation.

When horse traction gave way to the internal combustion engine the land was robbed of millions of tons of horse manure previously produced in town as well as country, and when modern methods of garbage disposal and water-borne sewage were introduced into our cities, the capital of the soil—its fertility—which is removed from it year by year in the form of crops and livestock, no longer found its way back to the land in the form of the waste products of the community, but was poured into the sea or otherwise destroyed.

[1] Howard, Speech at Crewe, 22 March 1939. Reprinted *New English Weekly*, 6 April 1939.

'On the basis of the data of Wolff, Kneller and Carpenter or of Hall, the people of the United States and of Europe are pouring into the sea, lakes and rivers, and into the underground waters, from 5,794,300 to 12,000,000 lb. of nitrogen, 1,881,900 to 4,151,000 lb. of potassium and 777,200 to 3,057,600 lb. of phosphorus per million of the adult population annually, and this waste we esteem one of the achievements of our civilization.'[1]

King, in his *Farmers of Forty Centuries*, writes: 'Man is the most extravagant accelerator of waste the world has ever endured. His withering blight has fallen upon every living thing within his reach, himself not excepted: and his besom of destruction has swept into the seas soil fertility, which only centuries of life could accumulate—fertility which is the substratum of all that is living.'

He points out that the Mongolian races with a population approaching 500 million, occupying an area 'little more than half of the United States, tilling less than 800,000 square miles of land and much of this during twenty, thirty or perhaps forty centuries, unable to avail themselves of mineral fertilizers, could not tolerate such waste and survive.'[2]

Nor shall we be able to survive indefinitely if we do not mend our ways.

The consequence of this process of denuding the soil of its fertility is only just beginning to be realized in Western countries. McCarrison has stated: 'These (certain natural foodstuffs), when properly combined in the diet, supply all the food essentials, known and unknown, discovered and undiscovered, needed for normal nutrition, provided they are produced on soil which is not impoverished, for if they be proceeds of impoverished soil, their quality will be poor and the health of those who eat them, man and his domestic animals, will suffer accordingly.'

Thus it will be seen that we cannot safely separate human health from the health of farm produce whether animal or vegetable. All have their origin in a fertile soil. Under field conditions a fertile soil is a live soil, and maintenance of life in such soil depends on humus.

This proposition is nowadays generally accepted in theory, but it is little understood in practice. Much of the existing confusion has arisen through a too frequent misuse of the word 'humus'. The term 'giving the land humus' is too often taken as a synonym for treating it with any form of organic matter, such as ploughing in green crops

[1] G. T. Wrench, *The Wheel of Health.*
[2] Cape, London, 1933, quoted Wrench.

or grass, or applying farmyard manure. But all these substances are merely some of the raw materials from which humus can be made. They cannot become humus until they have been metabolized by soil organisms. It is essential that this fact should never be lost sight of.

Humus is a product of decomposition of plant and animal residues, through the agency of micro-organisms. It has been simply yet accurately defined as the product of living matter *and* the source of it. 'The chemical composition of humus is determined by the nature of the residues from which it is formed, by the conditions of its decomposition, and by the extent to which it is decomposed. Chemically, humus consists of numerous organic complexes, the major group of which consists of lignins and lignin derivatives, and of proteins; a minor group contains carbohydrates, fats, organic acids, alcohols and other carbon compounds. The formation and mutual interrelations of these two groups of substances hold the key to the facts explaining the chemistry of humus.'[1]

Under natural conditions such as obtain in virgin forest or prairie, a constant supply of raw material is provided by the native flora and fauna which live and die, so to speak, *in situ*. Their residues and remains are continually converted by the soil organisms into humus from which fresh life can spring. A perfect balance between growth and decay is established, and the fertility of the soil is permanently maintained.

In ordinary farm cultivation we have a very different picture. Here, the vegetable and animal life raised from the soil is continually, and more or less permanently, removed from the site of its origin. If this process is to continue, fertility can only be maintained in one of two ways; either by supplying large quantities of organic raw materials from which humus can subsequently be manufactured in the soil itself, or else by deliberately manufacturing humus *outside* the soil, and applying it to the land as a finished product.

Ploughing in green crops and applying farmyard manure are methods of accomplishing the first of these alternatives, but they do not of themselves supply humus. The higher the yields required of a crop, the more advisable does it become to convert the available raw materials into humus before they are applied to the soil. For natural humus formation in soil is a slow process, and heavy cropping tends to exhaust the supplies faster than they can be replenished; moreover, raw organic matter, which is mainly cellulose, cannot become fully

[1] S. A. Waksman, *Humus*, Baillière, Tindall & Cox, London, 1937.

available to plants until it has become humus, and the organisms responsible for the conversion, require, for the process of manufacture, many of the elements also needed by plants.

'Humus is a manufactured product with a carbon:nitrogen ratio of about 10:1, prepared from vegetable and animal wastes with a carbon:nitrogen ratio of about 33:1. The conversion, which is carried out by fungi and bacteria, is naturally accompanied by the evolution of large volumes of carbon dioxide and requires a corresponding amount of atmospheric oxygen. Besides air, the organisms concerned in the manufacture also need water, a base to neutralize excessive acidity, and sufficient minerals, particularly combined nitrogen. Their demands are almost identical with those of the roots of plants. It follows, therefore, that any attempt to prepare humus in the soil itself is almost certain to interfere with the crop. Hence the injurious effects on growth which almost invariably follow the addition of straw, and very frequently, of green manure to the soil.[1] In such cases, the decomposition of these materials impoverishes the soil solution, contaminates the soil atmosphere, and often depletes the soil moisture. The result is that the soil is overworked and a poor crop is obtained. This overwork can be avoided by taking care to prepare humus outside the field rather than in the soil itself, and to restrict green manuring to localities where all these factors can be relied upon to yield a satisfactory result.[2] The Chinese were the first to grasp and act upon the master idea that the growth of a crop involves two separate processes: (a) the preparation of humus from vegetable, animal and human wastes, which must be done outside the field, and (b) the growing of the crop.'[3]

There are districts in China where the Chinese peasant, by adhering to this principle, has intensively and continuously cropped his soil without loss of fertility for forty centuries.

By 'humus manufacture' is meant deliberately speeding up the normal rate of cellulose decomposition. The process by which this is achieved is called 'composting'.

Composting is as old as agriculture. There is some evidence that many of the ancient civilizations, when in their prime, based their cultivation on systems of composting and irrigation, but whereas the

[[1] A solution to this problem is that discovered by Dr. Dhar of Allahabad University. See p. 14.]

[2] As for instance, in biologically active soil where the ploughing in is done in July or August between the spring and autumn nitrogen cycles.

[3] Sir A. Howard, *Manufacture of Humus by the Indore Process*, Royal Society of Arts, 1935.

primitive methods were, presumably, devised empirically, in recent years the knowledge contributed by biology, mycology and biochemistry has enabled a much more scientific approach to be made to the whole question of humus manufacture, and very exact and greatly improved methods of composting have resulted. While different methods vary as to certain factors and details of procedure, one fundamental principle underlies all successful compost making. Decomposition must take place by fermentation, and not by putrefaction. This involves careful control of the air supply, moisture content, and temperature developed in the compost heaps. Where these points are not properly attended to, the process has no right to be called composting, and the resultant product is not humus. It would be tedious in a book of this kind to describe in detail the different methods of composting in current use. To those who would like to make a comparative study of them I recommend a booklet by F. H. Billington, called *Compost*,[1] which contains a summary of them and also an excellent bibliography showing where further information can be obtained.[2] My own preference is for the method devised by Sir Albert Howard in the course of his research work in India, and generally known as the 'Indore Process of Humus Manufacture'.

I have now had considerable experience of the excellent results of this system under widely differing conditions. Of all methods, it has seemed to me to be at once the most adaptable to all systems of farming or gardening, and the most foolproof. This is an important point, for no method is likely to be widely adopted by practical cultivators, unless it is capable of being reduced to a formula, which, while precise, must also be simple. Howard points out that:

'When the Indore process in its final form was devised between the years 1925 and 1930, the scientific approach to the utilization of crude organic matter left much to be desired. The chief organic residues—farmyard manure, green manure, vegetable residues, municipal wastes, sewage sludge, and crude sewage, were being studied separately, and not as parts of a single subject. Much of the scientific work had been done, but the various fragments were lying around in the literature very much like the materials in a builder's yard, before the building itself is erected. On the practical side difficulties were being experienced. The results of green manuring were erratic; most of the methods of managing agricultural residues resulted in the loss of valuable nitrogen; some were elaborate, and some expensive; there

[1] Faber & Faber, London, 1942.

[[2] Apply also to the Soil Association. See p. 370.]

was no idea of examining the experience of old cultivation systems like the Chinese, the successful continuance of which over several thousand years is proof of efficiency, and illuminating them by the light of modern science.

'It is claimed that the Indore process has solved these difficulties. What was needed was the welding together of the separate fragments into a single well-ordered method, elastic enough to be introduced into any system of agriculture. In the course of its working out, the Indore process has been founded on correct biochemical principles, and is not far removed from Chinese or other more primitive practices evolved empirically in many parts of the world.'[1]

When you make Indore compost for the first time according to rule, it seems like a miracle that no matter what activator is used—pig muck, blood, night soil, or sewage sludge—the final product is inoffensive to crumble with the bare fingers, and has the pleasant earthy smell of old leaf mould.

[Note: Since this book was written composting is so generally, and so much better understood, and detailed instructions on how to go about it are so readily available that I have omitted from this edition the original description of the normal method of making Indore compost in heaps and pits. I have, however, retained the following section on sheet composting, as the best methods for carrying out this system are less well known.]

It is not always feasible to collect all vegetable wastes of a farm for composting in heaps. Some of these wastes will be crop residues still growing in the field, or else grass and clover leys to be ploughed up, or even crops specially grown for ploughing in as green manure. In these cases the operation known as *sheet composting* can be adopted. There are two methods of accomplishing this. The first is to dress such crops with farmyard manure during the summer while they are still growing (in the case of grass the best time is directly the first cut of hay is off), next allow the crop to grow up through the manure to a height of eight inches to one foot, then plough—not too deeply—while there is still enough warmth in the soil for decomposition to take place. September is the latest month in which this can be done. This method is very successful, complete decomposition taking place in good time for an autumn seeding.

The second method is essentially the same, but the manure is deposited on the crop by close grazing with livestock, instead of being made in the yards and carted on. When this method is used the stock

[1] See footnote [2], p. 66.

must be removed in time to allow the crop to grow to the requisite height before the August or September ploughing.

'One of the earliest and most effective examples of this process was originally devised by Mr. Hosier on the poor chalk pastures of Wiltshire. Following the introduction of the bail system,[1] the undecayed organic matter in the original turf was rapidly converted into humus by means of the urine and dung of the cow. The herbage improved until about the fifth year when no further progress was made. The turf was then ploughed up and converted into humus for the benefit of two or three straw crops, followed again by a temporary ley.'[2]

It is often asked whether the time and trouble involved in composting can be justified financially, so let us look at some of the advantages of the operation.

First and foremost must be placed the high degree of disease resistance in both crops and livestock that appears to follow liberal applications of well-made organic compost. Evidence on this point will be considered later. Second, the process enables all rubbish, wastes, and weeds of the farm to be utilized profitably. Third, the loss of nitrogen which usually occurs in the ordinary manure heap is avoided, as well as the bad smell and the formation of fly-breeding sites, and lastly, three times the bulk can be made from the same head of livestock, and there is evidence to show that the resultant product is superior. A large-scale test, involving forty fields, was made on the late Sir Bernard Greenwell's estate of humus manufactured by the Indore process versus best-quality farmyard manure, that is to say, well-rotted manure that was practically pure dung. Load for load the compost showed in every case better results.[3]

In 1942 I carried out a somewhat similar test to determine the relative merits, ton for ton, of Indore compost and *fresh* farmyard manure. The crops chosen were potatoes and brussels sprouts, followed, in 1943, by oats and barley respectively. In the case of the potatoes, compost at 6 tons per acre produced substantially the same result as farmyard manure at 12 tons per acre. In the case of the sprouts, where the compost and farmyard manure were both applied at the rate of 7 tons per acre the crop was markedly better on the composted portions. In both cases the 1943 cereal crop was better where compost had been used, showing that the previous year's

[1] Under this system the cows are kept in outdoor folds, the movable milking plant known as the bail following them round the fields.

[2] Howard, 'The Manufacture of Humus by the Indore Process', *Journal of the Ministry of Agriculture*, 1938.

[3] *Journal of the Farmers' Club*, 1939.

results were not merely due to plant food in the unrotted farmyard manure not being fully available to the first season's crop. These results, which fully endorse Sir Bernard Greenwell's findings, have an important bearing on the economic aspect of composting, for they indicate that the output, in farmyard manure, of a given head of livestock is capable of maintaining the fertility of two or three times as much land when it is made into compost as when it is applied to the soil direct, since the proportion of farmyard manure to vegetable rubbish in compost is one-third by volume.

Some mention must now be made of Howard's research work in India in the course of which the Indore process was evolved. Just as McCarrison succeeded in preventing disease among his rats by diet alone, so did Howard succeed in the prevention of disease among his crops and livestock by treating the soil with manufactured humus. In fact a very significant parallel exists between the experimental work of these two men.

Although their investigations were carried out quite independently, one in the medical and the other in the agricultural field, these can now be seen to have been complementary, forming two parts of a connected whole.

Both McCarrison and Howard began their research work on nutrition from the conventional approach of pathology. Both came to see that in order to discover the secret of health it is necessary to study the healthy and not the diseased organism. Both had the good fortune to be provided at a critical moment in their careers with ample facilities for untrammelled research. McCarrison, when he was appointed Director of Nutrition Research in India and given his laboratory at Coonoor, and Howard when he was appointed Imperial Chemical Botanist to the Government of India at Pusa. Here, he tells us, he was for the first time provided with 'real facilities for work; land, money, and freedom to grow crops in my own way, and to observe among other things the reaction, to insect and fungus pests, of suitable varieties when properly grown. My real education as an investigator then began—six years after taking my degree and after obtaining all the paper qualifications then needed for research. My duties at the Pusa Research Institute, fortunately for me, had not been clearly defined and I escaped the fate of many of our agricultural investigators—a life devoted to a research organization already becoming obsolete. It was possible, therefore, to attempt to break new ground and to try out an idea which had occurred to me in the West Indies, namely, to see what happened when insect and fungus diseases were left alone and allowed to develop unchecked, and where indirect methods only, such as a combination of better varieties and improved methods of agriculture, were employed to prevent attack.'

He took up all the land that was still available at Pusa, some seventy-five acres, and spent his first five years in India ascertaining by practical experience the principles underlying health in crops. He found that his best teachers were the peasants of India themselves, and the insects and fungi which attack crops.

'By 1910,' he writes, 'I had learnt a great deal from my new instructors—how to grow healthy crops practically free from disease without any help from mycologists, entomologists, bacteriologists, agricultural chemists, statisticians, clearing-houses of information, artificial manures, spraying machines, insecticides, fungicides, germicides and all the expensive paraphernalia of the modern Experimental Station.'

In the course of the work which resulted in this very considerable achievement, Howard formed the opinion that the key to disease resistance was a fertile soil, and that soil fertility could not be brought about by the use of artificial fertilizers, but was absolutely dependent on adequate supplies of humus. His subsequent experiments convinced him that the best results are obtained when humus is scientifically manufactured outside the field, and that this humus is only fully effective if composed of both vegetable *and* animal waste products.

Before the Indore process was evolved, tests were made to see if the same results could be obtained without the compost containing any proportion of animal wastes; but the peculiar power of humus to give resistance to disease was markedly less when the animal element was omitted.

'Ill-advised attempts in Ceylon to make compost with vegetable wastes alone have resulted in failure for the investigator himself and hindrance and discouragement for the other planters who were tempted to follow his example. The actual growth following the application of compost made from vegetable wastes alone is definitely improved, but no resistance to disease is established—indeed, a reverse effect has been observed.

'On the other hand, I have before me two reports from estates in Africa where animals have lately been included in the estate economy solely for the purpose of making up the necessary complement of animal wastes in the compost heaps, with extraordinarily impressive results. Large-scale experiments will soon provide incontrovertible evidence on this point; but it has been clear to me since the inception of the Indore process that animal wastes, even if (as in certain places in Great Britain) such substitutes as dried blood are resorted to in the absence of farmyard manure, are absolutely necessary.'[1]

An explanation for the failure which usually results from the omission of the animal element will be given in Chapter 5.

Howard further reached the conclusion, as a result of his experiments, that crops have a natural power of resistance to infection, and that proper nutrition is all that is required to make this power operative. Here was a close parallel to the rats of Coonoor; but in comparing the work of McCarrison and Howard, a point of particular interest emerges. When McCarrison's rats reached an age comparable with the fifties in human life, those which had been fed on the sort of food common among English working people were found, you will remember, to suffer from the same *degenerative* diseases which are rife amongst our population, while complete freedom from such diseases was displayed by the rats fed on the food of Hunza or Sikh, but McCarrison did not, so far as I am aware, attempt to cure the sick rats by a return to the Sikh diet, nor does he appear to have tested the two groups for their comparative powers of resistance to *infective diseases*.

The results of Howard's work on the other hand suggest that crops and livestock raised on land made fertile by his methods of humus treatment attain a high measure of immunity from infective and parasitic, *as well* as from degenerative diseases. Further, it would appear that this treatment is curative as well as preventative.[2]

In three centres in India, Howard was in control of extensive areas of land for a period of twenty-one years, and there he was able to carry out tests to prove his theories.

He claims to have found that the factor 'that matters most in soil management is a regular supply of freshly made humus, prepared from animal and vegetable wastes and that the maintenance of soil fertility is the basis of health'.[3] He further claims that his crops grown on land so treated resisted all the pests which were rife in the district and that this resistance was passed on to the livestock when these were fed on crops so grown.

'I then took steps to have my own oxen and to ascertain from first-hand experience the reaction of well-chosen and well-fed animals to diseases like rinderpest, Johne's disease and so forth which are

[1] Howard, *Humus*, Revista del Instituto de defensa del Cape de Costa Rica, 1939.

[2] See Chapter 6.

[3] Howard, quoted from 'Medical Testament'.

common in India. After a short time my animals duly came into contact with other oxen suffering, among other things, from foot-and-mouth disease.

'I have myself seen my oxen rubbing noses with foot-and-mouth cases. Nothing happened. The healthy, well-fed animal reacted towards this disease exactly as improved and properly cultivated crops did to insects and fungi—no infection occurred.

'These preliminary results suggested that the birthright of every crop and of every animal is health and that the correct method of dealing with disease is not to destroy the parasite but to make use of it for keeping agricultural practice up to the mark—in other words to regard the diseases of crops and livestock as Nature's Professors of Agriculture. These ideas were put to the test during the next twenty-one years at three centres in India, at all of which I had to manage large areas of land, and look after numerous oxen: Pusa (1910–24); Quetta (summers of 1910–18); Indore (1924–31). Everything possible was done to grow crops properly; everything possible was done for the livestock as regards food, hygiene and general management. The result was freedom from disease.'[1]

These remarkable results are by no means isolated.

The Indore process has since been put to the test in other parts of India, and also in many widely separated places throughout the world, including this country. Its adoption has been attended by remarkable, and sometimes even spectacular, results.

Representative examples of these results will be given in a later chapter, but first, in order that you may be in a position to form an unbiased judgement on the value of the evidence that can be deduced from them, it is necessary that the true nature of compost action should be made clear. Its behaviour differs widely from that of an ordinary manure, and it is important that this fact should be grasped, and the reasons for it understood.

Compulsory Pasteurization

I have omitted the original technical reference, which was a statement on pros and cons, because pasteurization of all milk, for all practical purposes, now being compulsory, the argument is only of academic interest, but perhaps it is useful to recount here that at a time when the only water supply on my farm was an open moat, I entered a county clean milk competition. Despite having only this

[1] Howard, quoted from 'Medical Testament'.

moat water for washing the cows and the dairy untensils, and only a home-made sterilizer, I not only won the competition, but my average bacterial count per c.c. was under 5,000, and sometimes under 500 (the maximum allowed for certified milk—the highest grade—is 30,000). For many years I sold my milk retail in a town over fifteen miles away and guaranteed its keeping quality for forty-eight hours in the hottest weather.

It is still possible for some country folk to get unpasteurized milk legally, but in towns and cities there is no longer any source of supply. This is to be regretted. With the high standard of hygiene which all producers of milk for sale must now follow, and with the ease of refrigerated transport, there is much to be said for having two classes of milk—'cooking milk' which might just as well be pasteurized, because it is going to be heated anyway, and 'drinking milk' (for those who wanted it) which would be raw milk, supplied under licence from certified healthy cows. Nutritionally there is a world of difference between the two. Pasteurization, and also of course cooking, injures both the vitamin and calcium content of milk, and the beneficial lactic acid bacteria, which attack disease organisms (and in the process cause souring) are killed by heat, leaving pasteurized milk as a perfect breeding ground for any subsequent contamination. [The most telling demonstration of the nutritional difference between raw and heat-treated milk (by whatever method) was the late Dr. Pottinger's famous experiments, a short account of which will be found in Appendix 11 at the end of Part II.]

4

Direct Evidence[1]

[Although this chapter no longer contains the photographic plates which were especially helpful to non-technical readers, the sources of the original illustrated reports are given in every case, and the most important of these—various issues of *Forestry*, the journal of the Society of Foresters of Great Britain—must, I feel, still be available in the Society's archives, for anyone wishing to study them. Most of the references originally supplied for this chapter (consisting mainly of tables of figures of weights and chemical analyses) I have omitted from this edition.]

I have stated that Howard and others claim to have proved by practical field tests extending over a number of years, that humus treatment confers on plants a power of disease resistance amounting in some cases almost to immunity, and that a like result is not, and cannot be, obtained by the use of artificial fertilizers. These views, however, are by no means universally accepted; it is therefore necessary to examine in some detail the scientific evidence in support of them.

As this assumes the probability that the soil micro-flora plays a direct and important part in the nutrition of the higher plants, I must first present the evidence on which this assumption is based.

This is an extremely complex subject, and a wide diversity of views exist concerning it, many of them still controversial. That soil fungi are capable of utilizing complex organic substances not directly available to higher plants, and of breaking these down into simpler forms, is an established fact. That many of these fungi also excrete substances that appear to act as a stimulant to growth also seems clear from the evidence. But the now very considerable and growing body of research work on mycorrhizal association seems to suggest that very

[1] This chapter, and the one following it, are very important to the argument of the book, but it is not essential that they should be read by the non-scientifically minded, or by those who are only interested in the conclusions to be drawn from them. Any reader, therefore, who finds them stiff, will lose little by reading the rest of the book first. This applies specially to Chapter 4.

many plant species and families benefit from the presence of specific soil fungi in a particular and intimate way.

The history of this research goes back nearly 100 years. Thread-like structures within the root cells of certain vascular plants were noted as early as 1829, and by 1847 were identified as being fungus mycelium. Frank described the 'dual structure formed by the tree root with its associated mycelium' under the name of *Mycorrhiza*[1] and the name has stuck.

'The outstanding results of Frank's preliminary investigations were, firstly, the recognition of root infection as a widespread phenomenon in trees, and the bestowal upon it of a distinctive name marking its existence as a morphological entity; and, secondly, the rejection of the accepted view of parasitic invasion of these roots, whether by truffles or other soil fungi, and the substitution of his theory of a symbiotic relationship beneficial to the trees. The far-reaching character of this hypothesis was a direct incentive towards the collection and interpretation of new facts bearing on the subject.'[2]

This collection and interpretation has gone on ever since. It is a fascinating story, and has been fascinatingly told by Dr. M. C. Rayner in her monograph on the subject.[2]

Since ignorance concerning this aspect of plant nutrition is very widespread, it may be helpful to give a very brief description of the mycorrhizal habit.

Mycorrhizas are usually classified, according to their structure, into two main groups, called ectotrophic and endotrophic, though these structural types are not nowadays considered to be sharply marked off one from the other.

The mycorrhizas of many trees and shrubs belong to the former group. These can easily be recognized by their outward appearance as distinct from ordinary rootlets. They form the short root system of such trees as pines and spruces, and members of the oak and beech family. A sheath of mycelium is formed over the tip of the young rootlets, and the internal fungus infection is mainly *inter*-cellular.

Members of the second group resemble ordinary roots to the inexperienced observer, though it is often possible to identify them by irregularity in diameter and differences in opacity. The fungus

[1] From *Myko* = fungal, and *Rhiza* = root. Should strictly only be used to denote a true fungus root, but is often loosely used to describe all mycorrhizal infection.

[2] M. C. Rayner, *Mycorrhiza*, Cambridge University Press, 1927.

infection of this group is mainly *intra*cellular, the mycelium within the cell walls eventually undergoing complete digestion by the plant cell. This process is plainly visible under a microscope. It is to this group that the mycorrhizas of the majority of crop plants belong.

The function of mycorrhizal association has been, and to a certain extent still is, a matter of controversy. The views of the early investigators ranged from the theory that the association was an accidental parasitic attack injuring the roots, to that of a mutually beneficent symbiosis, playing an indispensable part in plant nutrition.

The effect of the most recent research is to discredit entirely the first view. A phenomenon can hardly be accidental that occurs 'in wild and cultivated plants; in those growing in tropical forests and within the Arctic Circle; in habitats so different as the high Alps and the salt marsh, affecting an immense number of species of diverse families and occurring regularly under the most varied conditions of climate and soil.'[1] The list of known mycorrhiza-formers is being continually extended; at present it includes among tropical and subtropical crop plants, tea, coffee, sugar-cane, oil palm, coconut, cocoa, rubber, and tobacco; and among those grown in these islands and Europe, grasses including the cereals, potatoes, vine, hops, clovers, peas, beans and other leguminous crops, as well as many wild plants. Truly 'the habit challenges attention as a factor of significant potential importance to practical growers'.[1]

The theory of parasitism is equally untenable. In some highly specialized groups such as some of the orchids and heaths, the mutualism is so clearly defined that in some cases normal seedling development depends, under natural conditions, on the presence of the fungus.[2] In others the host plant dies or suffers from completely arrested development without its fungus associate.[3] In other less specialized varieties, it has been noted, among all those that have been closely studied under natural conditions, that the healthiest, most vigorous plants are those with the most active and widespread fungus infection, whereas in poorly developed and unhealthy plants, infection is weak and localized or even absent, so that to accept the theory of parasitism 'involves the paradoxical assumption that healthy growth is invariably accompanied by decreased resistance to

[1] Rayner, 'Mycorrhizal Association in Crop Plants', *Empire Cotton Growing Review*.

[2] For example Ling or Scots Heather (*Calluna Vulgaris*). See Rayner, *Mycorrhiza*, p. 96.

[3] See below. Also experiments with beech described in *Mycorrhiza*, p. 24.

parasitic attack, and by the production of an absorptive system the major part of which is hampered by fungal interference'.[1]

Since vigorous fungal activity in soil is dependent on adequate supplies of humus, it should now be clear that some knowledge of the part played in the life of the higher plants by soil microflora and fauna in general, and mycorrhizal association in particular, is an indispensable equipment of any would-be entrant into the arena of the Humus *v.* Chemical controversy, yet it is my experience that the majority, if not all, of the disinterested advocates of chemical fertilizers are totally ignorant of the subject, and that their ignorance is shared by many of their opponents in the humus camp.

This ignorance is perhaps excusable when it is remembered that even today few botanical textbooks give an adequate or unbiased account of mycorrhizal association,[2] and still fewer indicate the potential practical importance of this widespread habit among vascular plants.

Indeed the extent to which the whole subject has been ignored, even by accepted experts in soil science and plant physiology is difficult to understand. It is probably due to a variety of causes, of which the principal one is, I think, the over-emphasis on chemistry which, since Liebig, has been a feature of agricultural and horticultural research. This has resulted in soil chemists rather than soil biologists being placed in control of experimental stations. Secondly, soil biologists have tended to focus their attention on *specialized* groups of soil organisms such as, for example, nitrogen-fixing bacteria, and specific parasitic fungi. They have largely overlooked the general relationship of the soil micro-flora with *normal* plant growth.

This fragmentation, while understandable, nevertheless seems to me to be the curse of much modern research. Whatever we study, our tendency is to break it up into little bits, thereby destroying the whole, and then to study the effect or behaviour of the separate pieces as though they were independent, instead of—as in fact they are—interdependent.

The importance of the science of ecology in all fields of living matter has been increasingly realized in recent years, and it is recognized that the study of plant ecology is still in its infancy. Using the same analogy one might justly say that the study of soil ecology is in the pre-natal state. There are, however, signs that the birth pangs

[1] Rayner, *Forestry*, 1939.

[2] An exception is M. Skene, *Biology of Flowering Plants*, Sidgwick & Jackson, London, 1924.

have begun, and the infant, when it arrives, promises to be a lusty one, destined perhaps to play a vital part in man's better understanding of how to live.

When this happens no research worker will have a greater claim to rank among the chief assistants at its birth than Dr. Rayner. Her recent work on mycorrhizal association has been mainly concerned with trees, and her investigations are quite separate from, and independent of, any of the aspects of soil fertility and health dealt with in this book. Their importance in connection with this subject rests mainly on the fact that an incidental result of her work has been greatly to elucidate the nature of compost action, and the effects of its application on the soil population.

The exact manner in which associated or other soil fungi benefit higher plants is a question on which experts still differ. I discuss some of the different theories in the next chapter. The accurate answer to the question is of immense scientific interest but from the practical point of view it is more important to establish the fact that the plant *does* benefit, and in a way that is vital, either directly from the presence of these fungi, or from soil conditions which also favour free fungal development, for obviously if such a fact were fully established in regard to crop plants the implications would be very far-reaching. Among other results would be a complete justification of Howard's conclusions, for, as already stated, fungi thrive in humus and cannot live in soil without it; moreover there are indications that soil fungi are inhibited by many chemical fertilizers.[1]

The difficulty of providing critical proof of the beneficent action of soil fungi on crop plants arises from the paucity of direct scientific evidence regarding the physiological role of endotrophic groups of mycorrhizal fungi in plant nutrition, from the scattered nature of the circumstantial evidence, and from the extreme complexity of the subject. This last may well be responsible for the different results sometimes obtained by different investigators concerning the relative merits of organic versus chemical soil treatments.

My difficulty, therefore, in adventuring into this sea of cross-currents was to find a really secure anchorage from which to start. When, in the course of extensive reading on the subject, I obtained access to the reports of Dr. Rayner's recent work on trees at Wareham Forest, I decided, rightly or wrongly, that I had found such a starting point.

True, these investigations were concerned with the ectotrophic

[1] See Chapters 5 and 6.

mycorrhizal responses in coniferous trees growing on natural soils, and might at first sight appear to be totally inapplicable to crop cultivation, nor do I claim that they are *directly* so applicable. The great advantage of them, as a starting point from which to consider the evidence in a wider field, is that while most of this other evidence is at best only circumstantial, though none the less cogent, the evidence of Rayner's experiments provide, in the case of certain important conclusions, critical scientific proof of so definite a nature as to preclude the possibility of doubt concerning the correctness of the deductions.

This being so, I felt that despite the specialized nature of the material under investigation at Wareham, here was a yardstick by which some of the scattered evidence covering a much wider field of material and conditions might perhaps be measured. My reasoning ran thus: in the course of Rayner's investigations on pines, certain soil treatments produced certain responses in the tree seedlings. The reasons for the results obtained were definitely and fully established. If it could be shown that similar treatments produced similar responses in crop plants on cultivated soils, then a prima facie case would have been made out for assuming that in both cases the same cause was operating.

I received early encouragement to pursue further this line of thought when I learned that the indigenous vegetation growing in the experimental area at Wareham did in fact respond to certain treatments in much the same way as the trees.

In the next chapter I shall develop further the hypothesis that an analogy exists between the Wareham experiments and crop cultivation. First I wish to describe two of the experiments themselves. The first proved that compost does not act like an ordinary manure by supplying nutrients to the pine roots direct (although this discovery when originally made was quite incidental to the work in hand), but that it acts indirectly by stimulating the activity of soil fungi. The second proved that the action of these fungi, in association with the pine roots, stimulated both root and shoot development in a manner far surpassing anything obtained by inorganic chemicals, this being demonstrated not only by the size of the tree seedlings under investigation, but also by their health and vigour, and powers of disease resistance.

I am greatly indebted to Dr. Rayner, and to the editor of *Forestry* for permission to use abstracts and quotations from the papers describing these experiments.

Dr. Rayner is perhaps the first botanist to realize that the question of the exact function of mycorrhizal association in trees, so long a matter of controversy, can never be finally answered in the laboratory.

'Modern research on mycorrhizal problems has been marked by recognition of the extraordinary frequency of the phenomenon in nature and by the application of more precise methods of research involving isolation and pure culture of the fungi concerned. Investigations of this kind are laborious; extraction of the mycorrhizal fungi and the synthesis of fungus and host in pure culture present difficulties, while evidence of behaviour under fully controlled conditions must be applied with the greatest caution to those in nature. The mycorrhizal habit, possibly of critical importance to plants exposed to full competition in the field, may play an insignificant part in the nutrition of the same species sheltered from such competition. For the latter, the rooting conditions are relatively simple and static; to the former they must present a succession of complex and continually shifting problems complicated by the direct competition of other soil organisms for food materials, and also indirectly by changes in the environment conditioned by these and other microbiological activities.'[1]

Holding these views, it was only natural that Rayner should be on the look-out for opportunities of conducting research on mycorrhizal reaction under natural conditions, and of supplementing laboratory tests by field tests.

An admirable opportunity occurred in connection with the afforestation under the Forestry Commission of a tract of 3,454 acres of very poor infertile land at Wareham Forest in Dorset.

The early attempts at tree planting there had met with very irregular results, a large proportion of the seedlings having died outright or passed into a condition of 'more or less complete check'. Rayner records that 'It is still possible 14–16 years after these original sowings, to find surviving seedlings a few inches high. The root systems of such plants from untreated soil even when quite young, were defective, with intermittent die back of the younger roots, and corresponding poverty of mycorrhizas.'[1]

In 1930 Rayner began her series of researches into the cause of infertility for tree growth in the Wareham soil. She formed the opinion that in those places where tree growth was inhibited a toxic factor in the soil was operating. This view was subsequently proved to be correct by independent investigation and the toxin was shown to be

[1] Rayner, *Forestry*, 1939.

of biological origin.[1] This peculiarity in the Wareham soil greatly facilitated the investigation. One of the factors adding to the difficulty of mycorrhizal research under natural conditions is the liability of roots to suffer attack by parasitic fungi, when conditions favouring mycorrhizal association are upset. In untreated Wareham soil *all* fungal growth is more or less inhibited, and thus experiments designed to determine mycorrhizal responses were exceptionally free from any confusing issues.[2]

The earlier experiments at Wareham proved conclusively that mycorrhizal association in pine is causally related with healthy growth, and experimental soil treatments were designed to ameliorate the conditions believed to be responsible for inhibition of fungal activity, and ultimately for defective root growth of the pines sown or planted in the area.

Chief among the treatments adopted by Rayner to this end was the use of different kinds of carefully prepared organic composts. The results were startling. They were published in the 1936 issue of *Forestry*, and caused a flutter in forestry circles; also some controversy, for while the results of her treatment could not be disputed—they were there for all to see—her conclusions were not always accepted. At that time proof was still lacking of the correctness of her initial hypothesis of the biological origin of soil infertility at Wareham, and it was argued that it was the chemical nutrients in the composts that affected the growth of the trees, and not their effect upon biological soil activities. An experiment was therefore designed as an answer to this criticism.

Field observations had already indicated the complex nature of compost action. For example, tests were made with two composts of almost exactly similar appearance and chemical composition. One was made with 75 per cent softwood sawdust and 25 per cent fallen beech leaves, and the other with 75 per cent softwood sawdust and 25 per cent chopped straw, both were composted with 1 per cent of nitrogen in the form of dried blood. When these were applied to field sowings of Scots pine, the first produced 'well-balanced, vigorous growth with free development of mycorrhizas of normal structure', while the second produced results 'little better than the untreated controls'. Yet the same two composts when applied to another species

[1] Professor Neilson-Jones, 'Biological Aspects of Soil Fertility', *Journal of Agricultural Research*, XXXI, 4, October 1941. (See also Chapter 5 and Tech. Ref. No. 2, p. 89.)

[2] See Tech. Ref. No. 2, p. 89.

of pine (*Pinus Contorta*) both produced equally vigorous growth, in marked contrast to the controls.

The whole of the experimental area suffers from a marked phosphate deficiency, and the application of phosphate in certain forms (notably basic slag and bone meal) has brought about effects in kind similar to that induced by compost treatment, though the *extent* of growth stimulation is greater with composts. On the other hand phosphate applied in some other forms, superphosphate for instance, proved lethal to the trees.

These various observations are held to support the view that compost, and very probably phosphate also when applied in the form of basic slag or bone meal, act indirectly, and that their organic constitution, and their effect on the soil micro-flora, are more critically important to the growth of conifers, than the quantity of available nutrients which they may contain. In order to put this hypothesis to the test the following experiment was carried out.[1]

Several series of pot sowings of Scots pine were made in imported Wareham soil. In two series, two different composts (C 1 and C 5) were added at the time of sowing. Two other series received the amount of salts needed to supply nitrogen, potash, phosphoric acid, and lime, in the same quantities as those present in the compost. A fifth series was sown in the experimental soil with cow manure added at the same rate as the compost. The control series consisted of sowings in the experimental soil without any addition.

C 5 compost is made from hop waste, and C 1 compost from chopped straw. In this experiment both were composted with 1 per cent nitrogen in the form of dried blood.

All series were run in two alternative ways. In one group any excess of water which drained from the pots was discarded. In the other it was collected in darkened containers, in which the pots stood, and was regularly returned again to the pots. In this second group, therefore, no soluble nutrients were lost through leaching. The plants were watered with rainwater throughout the experiment which lasted for two years.

Rayner reports that all cultures were 'entirely satisfactory in respect to health of the seedlings, uniformity of growth within each series, and range of variation within the populations of individual pots.'

At the end of the first year's growth control seedlings in the untreated Wareham soil 'made defective root and shoot growth and

[1] See *Forestry*, 1939 issue, or M. C. Rayner, *Problems in Tree Nutrition*, Faber & Faber, London, 1944.

produced short needles of a bad colour'. The series receiving equivalent salts were 'quite healthy and vigorous' but were markedly smaller than the plants of either composted series. Both these last made much more growth than any of the other series, the maximum being obtained with C 5 compost. In both, 'shoot and root development were greater, needles longer and of better colour'.

There was very little difference in any of the series between the two groups, one with and one without having the drainage water returned.

During the second season the differences in growth between the various series was maintained. But by the end of that year when the experiment was brought to an end, a marked difference had become apparent within the different series between the groups receiving and not receiving the return drainage water. This affected not only the size but also the health and vigour of the seedlings.

In the case of the compost treated plants, and those with cow manure, the return of drainage water proved to be beneficial, the dry weights of the seedlings in these series, exceeding those of the same series in the group not receiving the drainage water by from 2·2 per cent in the case of cow manure up to 12·21 per cent in the case of C 5 compost. In the case of the control seedlings in untreated soil and those receiving equivalent salts, the reverse effect was apparent. Here return of the drainage water proved to be deleterious. In this case the dry weights of the seedlings in these series were *less* than the same series in the group not receiving the return drainage water, by amounts ranging from 23·7 per cent in the case of C 5 salts up to 32·5 per cent in the case of the controls.

A further interesting point was noted, the increases and decreases in the dry weights were calculated separately for the root and shoot systems and it was found that where the return of the drainage was beneficial, the increase in dry weight was greatest in the case of the root system, whereas in those series where the return drainage proved to be deleterious the decreases in dry weights were greatest in the shoot system.

[The detailed tables of weights have not been included in the 1975 edition. Tables 1 and 2 give a summary.]

TABLE 1

WHERE RETURN DRAINAGE WATER WAS BENEFICIAL

	Increase weight percentages	
	Shoot	*Root*
Wareham Soil + C 5 compost	6·5	20·2
Wareham Soil + C 1 compost	0·7	20·3
Wareham Soil + Cow Manure	1·8	3·4

TABLE 2

WHERE RETURN DRAINAGE WATER WAS DELETERIOUS

	Decrease weight percentages	
	Shoot	*Root*
Wareham Soil untreated	37·3	20·0
Wareham Soil + C 5 equivalent salts	25·8	17·8
Wareham Soil + C 1 equivalent salts	27·3	24·5

Rayner's observations on this result are as follows:

'The presence of soil substances actively inimical to root growth was inferred at an early stage of the Wareham researches. It was, indeed, the conclusion derived from this inference, that the marked infertility of this soil for tree growth is more closely related to the presence of such toxic substances than to poverty of nutrients, that led originally to the use of organic composts as a means of alleviating the unfavourable condition.

'In view of the purpose of the present experiments, however, it is regarded as highly significant that return of drainage water over considerable periods has provided evidence of a depressing effect upon growth, far exceeding in magnitude any beneficial effect derived from return of leached nutrients, in all cases except those in which the soil received additions of organic matter in the form of composts or cow manure. The results of the present experiments thus confirm conclusions already reached in respect to the existence of deleterious substances in the soil solution, and clarify previous observations on the beneficial effects of leaching in pot cultures. Not only so, but they afford convincing evidence in support of the hypothesis that compost treatments operate mainly by bringing about fundamental changes in the organic residues, whereby the microbiological activities of the soil and their ultimate by-products are profoundly and permanently modified.

'The experiments show that increase in the supply of available nutrients plays a relatively insignificant role in the steady maintenance of healthy and vigorous growth that follows addition of composts to the soil. They show further that addition of mineral salts is practically without influence on the production of substances deleterious to growth, whereas that of composts—and to a less extent of cow manure—puts a stop to the production of undesirable by-products, thereby rendering the leachings innocuous and permitting utilization of any nutrients returned with them.'

Examination of the root systems for mycorrhizal reaction at the end of the experimental period showed that in these pot experiments, 'All plants showing maximum size, health and vigour, showed also a maximum development of mycorrhizal fungi.'[1]

Discussing the results of this experiment Rayner writes:

'The experiments described in this paper were designed to meet the criticism that addition of mineral nutrients to the soil, in amounts corresponding to those in composts, would produce similar effects on growth, and to establish the correctness of conclusions put forward in an earlier communication. For the soil under investigation, these conclusions were that the use of organic composts produced qualitative changes in the humus constituents and that these, rather than increase in the supply of available nutrients, are the fundamental cause of restored fertility. The failure of growth and irregularity of behaviour observed in the trees were interpreted as bound up with biological activities yielding inimical by-products, and not merely as a resultance of poverty in available nutrients.

'The evidence derived from the pot cultures in Series 2, 3, 4 and 5 of the present experiments affords direct support for this hypothesis of compost action. It proves that increasing the supply of nutrients in this poor soil by addition of soluble salts of nitrogen, potash and phosphoric acid, corresponding in amounts to those in the composts, evokes surprisingly little response in pine seedlings, even when potential loss by leaching is compensated for by return of all drainage throughout the period of the experiment. It thus provides a complete refutation of the criticism that the action of the composts in promoting the healthy growth of young trees on certain soils is mainly due to increased supply of directly available nutrients, confirming the

[1] Space does not permit the inclusion of a detailed description of the mycorrhizal responses in the different series of this experiment. Readers are referred to Dr. Rayner's paper in the 1939 issue of *Forestry* which is published as a reprint by the Oxford University Press, or to her book *Problems in Tree Nutrition*.

conclusion that the action of the compost in stimulating the growth of pine seedlings differs essentially from that of ordinary manures. . . .'

I have described this experiment in some detail, because the point which it emphasizes cannot be too often repeated, namely, that the value of compost as a treatment for crops cannot be measured only, or even mainly, in terms of its chemical composition.

It may be argued that no such general deduction can be made from the results of experiments carried out on conifers, and on a soil type so extreme as that at Wareham. I offer my answer to this possible criticism in the next chapter, when the whole question of the wider application of the Wareham experiments will be considered. At this point I go no farther than to assert that in so far as the pine seedlings under consideration were concerned, the experiment just described definitely established the claim that compost does not act like an ordinary manure. Further, it demonstrated that while compost possesses the power to remove factors inhibiting fungal activity, continued application of nutrients in the form of chemical salts does not.

That mycorrhizal activity follows the application of compost was proved by examination of the root systems concerned. That this increased activity was due to stimulation of the fungi already present, and not to active mycelium introduced with the compost was proved by the fact that mycorrhizal activity was also found to follow applications of sterilized compost.

Although these vigorous mycorrhizas formed by healthy seedlings treated with compost show marked structural differences from those growing in untreated soil, there is no question of a different fungus being responsible. In natural Wareham soil only one species of fungus has ever been observed as forming association with Scots pine.

'All mycorrhizas, whether normal in structure or otherwise, are formed by association with *B. bovinus* identified in the case of each type by isolation of the mycelium and comparative study in pure culture.'

We now come to the second point. The series of experiments which established the fact that it was the stimulation of the mycorrhizal fungus, by whatever means induced, that was responsible for the restored fertility. Proof of this has been supplied by a series of experiments, the results of which are described in a paper by Rayner and Levisohn published in the 1941 issue of *Forestry*.[1] These experiments took the form of isolating the mycelium responsible at Ware-

[1] Reprinted in *Problems in Tree Nutrition.*

ham for the formation of mycorrhizas in pine, and then using different forms of the pure fungus in a series of soil inoculations. Tests were also made with humus inoculations.

These investigations were carried out both in field plots and in pot experiments. In the latter, the sequence of events already observed in compost-treated plants followed in all cases, whether the inoculum employed consisted of pieces of natural sporophores, fragments of humus from suitable soil, or pure fungus culture raised on laboratory media. In all these cases inoculation was followed by development of the short root system of the seedlings, and the subsequent rapid conversion of these roots into mycorrhizas. A marked improvement in the health and growth of the seedlings then took place.

In the field-plot experiments the inocula were inserted here and there throughout the plots, a few inches below the surface. In this series of investigations the resulting responses were much more variable than in the pot experiments, but in all cases an improvement in growth as compared with the controls took place, though in some of the pure culture inoculations this was considerably delayed.

In these field-plot inoculation experiments, the improvement in the health and growth of the seedlings in the plots treated began with those growing nearest the points of inoculation, and gradually spread outwards from them, through the plot. Stimulus to growth must therefore 'originate in these inocula and is evidently biological in character. . . . It is legitimate to infer that the biological constituent concerned in all these cases is mycelium of a fungus or fungi known to form mycorrhizas with the tree species under observation; known to be introduced in the inoculum and observed to be present in the mycorrhizas subsequently developed. . . . The conclusion follows, therefore, that soil activities leading to the stimulation of plant growth are induced by the introduction to seed-beds of suitable inocula, whether of soil fragments or sporophore tissue, or pure cultures of specific mycorrhiza-formers,' and 'This stimulus to root production is directly related in some manner—at present unexplained—to the results of fungal activity in the soil.'[1]

As a result of the two series of investigations just described, it must be conceded that in the case of pine seedlings grown on Wareham soil, the following facts are fully established.

(a) Compost treatment produces healthy growth absent in the controls.

[1] For detailed technique of this experiment the reader is again referred to original sources mentioned in footnote to p. 80.

(b) This is not due to the available plant nutrients contained in the compost.

(c) Nutrients supplied in the form of inorganic salts fail to correct the toxic condition, and their continued application increases it.

(d) The vital factor governing restored fertility is stimulation of mycelial activity.

Before proceeding to the next step in the building up of my case there is one more of the Wareham results which must be mentioned. This is concerned with observations made on Lawson's cypress. This conifer, unlike pines and spruces, forms endotrophic mycorrhizas in common with ordinary crop plants. It is a native of the northern Pacific States of the USA and is commonly planted as an ornamental tree in gardens in this country. It is 'not fastidious in respect to soil and climate under these conditions'.[1]

During the earlier afforestation attempts at Wareham a few experimental plantings of this tree were made there. The seedlings used were transplants from nursery soil. Some were transplanted into untreated Wareham soil, others received 2 oz. of basic slag at the time of planting. When Dr. Rayner's attention was first drawn to these seedlings, four years after planting, they were 'conspicuous for the vigorous growth made on very poor light soil . . . both the control plants and those that had received phosphate were equally healthy and vigorous; in both the foliage was green, and there were no symptoms of malnutrition.'

This behaviour was so surprising in view of the natural habitat of this tree, and so at variance with the usual experience at Wareham, that an examination was undertaken of the root systems of seedlings growing at Wareham, of others in the original nursery soil, and of a third series sent from Oregon, USA.

In plants from all these sources 'the whole root system including the young active roots proved to be mycorrhizal, showing deep-seated infection by the mycelium of a fungus of the vesicular-arbuscular type'. Pot sowings were then undertaken in order to find out if this fungus was ordinarily present in Wareham soil, or had been introduced with the transplants. One series was sown in untreated soil, two in soil receiving varying amounts of basic slag, and another series in soil plus an addition of C 5 compost. Here are the results. The controls in untreated soil were 'very small with poor root development and discoloured foliage. Addition of basic slag at the rate of 5 grammes or more per pot was deleterious to growth; at the rate of 2 grammes

[1] Rayner, *Forestry*, 1941.

per pot produced no detectable effect. Addition of C 5 compost gave marked improvement of growth comparable with that already observed for pines and other conifers. In none of the experimental plants was any trace of mycorrhizal infection present.' This result was later confirmed by further experiments. The fungus had thus clearly been introduced with the transplants.

The sequel to the story is particularly interesting. In 1939 the cypress seedlings growing at Wareham began to show signs of ill health: first yellowing of the foliage and then cessation of growth. Laboratory examination of the roots of these plants revealed that 'mycorrhizal association had disappeared from the roots of the plants treated with basic slag, and was present only as traces of fungal infection locally in those of the control plants.'

Obviously mycorrhizal association plays a critical part in the normal nutrition of this tree. The two most interesting points in this investigation seem to me to be: (a) that while the fungal associate of this cypress (a forest tree in its native USA), clearly does not exist in Wareham soil, it does presumably exist in the garden soils of this country; and (b) that in the pot experiments the seedlings made healthy growth in the presence of compost even without forming mycorrhizal association.

In connection with this, an interesting thing occurred last winter (1941–2). Dr. Rayner tells me that in the course of a pot experiment with seedlings of this cypress, it was found necessary (not as part of the experiment) to let the pots stay out of doors all the winter. As most of us are not likely to forget, it proved to be the hardest winter known in this country for many years. The severe frost proved to be too much for the pot seedlings growing in untreated Wareham soil, or in soil with an addition of basic slag; the plants growing in compost-treated soil on the contrary, not only survived the winter, but remained outstandingly robust and healthy.

Tests are now being made at Wareham with this tree to determine whether applications of compost at the time of planting will maintain the activity of the associate fungus introduced with the transplants, and further tests are being made to discover whether this cypress can continue to thrive without its fungus if compost treatment is given. I await the results of these further experiments with lively interest, for I feel that Lawson's cypress, which, like crop plants, forms endotrophic mycorrhizas, may afford a link to bridge the gap in the present knowledge of the latter.

The knowledge which has been accumulated in respect to the

ectotrophic mycorrhizal habit in trees is now very extensive, as this chapter must have indicated. The endotrophic mycorrhizal habit in crop plants on the other hand has been shockingly neglected.[1] The fact that trees are long-term plants, so that seedlings can be observed over several years' growth has, no doubt, made them easier to study, and therefore more attractive to research workers. The ordinary crop plant which reaches maturity in a single season must obviously present greater difficulties to the would-be observer. This is where I feel that a study of Lawson's cypress may be of assistance. If this tree should prove capable of survival without mycorrhizal association in compost-treated soil, then a close study of its behaviour might lead to an explanation of the undoubted fact that on species of crop plants not believed to form mycorrhizal association,[2] compost treatment often produces results as spectacular as those resulting from the same treatment on known mycorrhiza formers.

This is an important point sometimes lost sight of. It will be more fully considered in the next chapter.

Technical Reference No. 1

(a) There is conclusive evidence that Wareham soil contains 'toxic' factors of biological origin that operate as inhibitors to fungal growth.

(b) In laboratory culture the inhibiting action has been directly observed upon a number of soil-borne and air-borne fungi, including proved mycorrhiza formers as well as other species that form pseudo-mycorrhizal associations with tree roots. So far as is known the effects produced are not lethal, but are evidenced as more or less complete inhibition of mycelial growth and of germination of spores. There is evidence of differential susceptibility among the fungal species observed.

(c) The condition of biological inertia thus brought about in the soil is manifested in many ways. It leads to the formation of a substrate unfavourable to root growth and affects mycorrhiza formation

[1] 'The part played in nutrition by the fungi present in endotrophic mycorrhizas, and the biological significance of root infection in relation to differences of soil and other external factors, continues to be a subject of controversy. There is a strong prima facie case for a beneficent action upon the nutrition of the host based upon cytological and microchemical evidence. To supplement this, data derived from pure culture experimental researches on certain groups is now slowly accumulating.' See also Tech. Refs., pp. 88–90.

[2] For example, members of the cabbage family.

directly by inhibiting mycelial activity, although suitable mycelium may be present in the neighbourhood of roots. Reduced fungal activity as compared with that in non-toxic soils of similar type is confirmed by direct microscopic examination; fungus mycelium is relatively scarce and there is a notable absence of the peritrophic fungus flora often conspicuous about roots and mycorrhizas. These characteristics greatly simplify observation of the mycorrhizal relationships in Wareham soil: there are present only few fungal species that form mycorrhizal association with the roots of pines and other conifers, and there is a notable absence of activity on the part of soil fungi known to cause parasitic or pseudomycorrhizal attack on roots; the task of isolating mycelia of the fungus associates from individual mycorrhizas is thereby made comparatively easy. (Rayner, *Forestry*, 1941.)

Technical Reference No. 2

ENDOTROPHIC MYCORRHIZA

(Extract from paper by E. J. Butler of the Agricultural Research Council published in the *Transactions* of the British Mycological Society, XXII, 3 and 4, 16 February 1939.)

'. . . the roots of most flowering plants and ferns, and the prothalli of the liverworts amongst the Bryophyta and of some of the Pteridophyta are subject to invasion by fungi belonging, with few exceptions, to a homogeneous type characterized by the possession of special organs, the vesicles and sporangioles. These fungi are not found in all roots and appear to be rare in many plants in particular localities. The evidence from cotton and the cereals shows, however, that absence or rarity of the fungus in one place is not inconsistent with prevalence in the same plant elsewhere. The association is, to a certain extent, casual, as has also been pointed out by Miss Ridler (1923) in *Lunularia*, but it need not on that account be without significance to either or both the partners in the association. Indeed the elaborate mechanism for the manufacture of fatty substances and their rendering to the higher plant can scarcely have no significance. With this mechanism must be associated the peculiarities of the special organs of the endophytes in question. The vesicles are constructed so as to store considerable volumes of an oily substance free to be reabsorbed back into the mycelium without the intervention of a septum to block the passage, and these stores of oil or fat are emptied into the root cells from the sporangioles. The latter are so designed as to offer the maximum of surface to the disintegrating action of the cell juices.

They are not merely artifacts due to a 'digestive' action of the cell on the fine branches of the arbuscules as appears to be suggested by some observers, but are definite bladders formed at the tips of these branches and passing through a stage of turgidity before they break down. Nor are they haustoria engaged in absorbing nutriment from the host cells as has also been suggested, for their formation often marks the end of the vegetative activity of the fungus in a particular region, or in a considerable area of fine roots simultaneously. That they have some function in the developmental nutrition of the higher plant, as most students of them have suggested, can hardly be doubted. The purely tentative hypothesis is advanced that this is not the supply of fat as such but of some fat-soluble growth or development-promoting accessory substance. The extraction of sufficient quantities of this hypothetical substance would present great difficulties so long as the endophyte defies artificial cultivation; but in certain plants such as bamboos or *Ricinus communis*, the volume of the endophyte may be very considerable in the white roots and it might be possible to obtain appreciable quantities of the fungus from them at an appropriate season.'

5
Circumstantial Evidence

[This chapter strikes me as no less important than it was when it was written. I, myself, had forgotten, until I re-read it, the strength of the evidence it presents for equating soil fertility with fungal activity and the importance of this evidence in interpreting the effect of composts. In recent years the vital role of soil bacteria in plant nutrition has become generally recognized, but it seems to me that the importance of fungi is still largely overlooked. The reprinting of this chapter I regard therefore as not merely of historical interest, but as both relevant for today, and timely.]

In this chapter I undertook to show that a connection exists between Dr. Rayner's research work on coniferous trees at Wareham Forest, and the everyday cultivation of food crops.

In order to marshal the evidence in an orderly fashion, I will now recapitulate under six headings certain facts proved at Wareham. I then propose to take these six points in order, and to see if any evidence exists for believing that to any, or all of them, a parallel can be found in ordinary farming practice.

Here is the list:

(a) Mycorrhizal association plays a critical part in the nutrition of pines, the fungal activity producing substances that stimulate the growth of lateral roots.

(b) Infertility for tree growth at Wareham is not due to absence of the appropriate associate fungus, but to a toxic condition which inhibits all mycelial growth. This toxin is of biological origin, being a by-product of the metabolism of some soil organism.

(c) The application of nutrient salts of nitrogen, phosphorus, and potash, fail to alleviate the toxic condition, the relatively small improvement in growth of the seedlings resulting from their application not being maintained.

(d) The application of composts made by fermenting vegetable wastes with 1 per cent of nitrogen in the form of dried blood, not only rapidly removes the toxic condition, but is self-propagating in that the soil so treated does not regain toxicity. This result is not obtained in so great degree when an inorganic chemical activator is substituted for the dried blood.

(e) The action of the compost is not due to the plant nutrients it contains, but to its biological reaction which has the effect of fundamentally modifying the soil microflora.

(f) The restoration of health and vigour in the pine seedlings, brought about by compost treatment, is not due to direct action of the compost on the plant roots, but to indirect action due to stimulation of mycorrhizal and other soil fungi, and the removal of the substance previously inhibiting their growth.

These six points were *proved* in so far as the particular case of tree growth in Wareham soil is concerned. On the extent to which they can now be shown to apply also to other soils and crops, depends their importance to farming practice, and national health.[1] Taking these points in order then, the first question to be answered is this: Does mycorrhizal association play a vital part in the nutrition of other mycorrhiza formers growing on ordinary soils? If so, in what does the benefit to the host plant consist?

The following is the opinion of a Professor of Botany on this point:

'The plant physiologist is simplifying unduly the problem with which a plant growing in soil is faced in obtaining nutrients by means of its roots, if it be assumed that conditions approximate to those of experimental laboratory cultures in which mineral salts are supplied to the roots for absorption. In soil, the activities of various microorganisms complicate the natural roots' environment in many ways: not only by direct competition but by controlling the forms in which available nutrients are presented and the production of substances beneficial or deleterious to growth; also, by forming beneficent or harmful associations with the root tissues. The fact that probably 80 per cent of flowering plants form mycorrhizal relationships is alone sufficient to suggest that the simple picture of mineral salts absorbed through root hairs represents only part of the mechanism of root nutrition—in many cases quite a small part of all the processes intervening between addition of organic matter to the soil and absorption of nutriment by the plant. Presumably, the difficulty of precise quantitative investigation of the nutritive relations of mycotrophic plants is responsible for retention of the convenient fiction that this form of nutrition is casual and exceptional, when in point of fact it

[1] See Chapter 6.

appears to be normal and regular for a majority of mature flowering plants in soil.'[1]

In a pamphlet, printed in 1939, of which this [original] book is really an enlarged and revised edition, I suggested that the part played by mycorrhizal fungi in forming a bridge between humus in the soil and the roots of plants, was a discovery of very recent origin.

It will be seen from the previous chapter that this is not factually correct, Frank having held the view, as long ago as 1885, that 'Mycorrhiza is a symbiotic relation to which probably all trees under certain conditions are subject. It is formed only on soils containing humus or abundant plant remains, and its formation waxes and wanes with the abundance or otherwise of these constituents in soil. The root fungi carry to the trees not only the necessary water and salts but also soluble organic material derived from the humus, thus lending a new significance to leaf-fall and the accumulation of humus in woodland soils.'[2]

But even though this discovery was made by 1885, I still do not feel inclined to retract my original statement, for while it may not be true in the letter, I maintain that it is true in the spirit. In support of this view I again refer to the inadequate treatment of the subject in textbooks, and agricultural teaching. Although research work on mycorrhizal responses has been going on for 100 years, it has been principally confined to trees, orchids, and heaths, and mainly conducted in the laboratory. That the possible importance of the habit in ordinary crop plants has been largely overlooked is clearly shown by the first of the above quotations. Not till Rayner's field experiments demonstrated the connection between compost treatment and mycorrhizal responses in trees, was the clue apparent to explain results being obtained, by some other users of compost, on crop plants.

The clue may have been on record for close on 100 years, but Rayner presented it in a new light, and Howard had the brainwave to see where it led.

How new was this light is proved by the fact that it was found necessary to arrange a special experiment in order to demonstrate that the beneficent effect of compost bears little relation to its chemical composition, and that its biological reactions are of more importance than either its physical properties or the plant nutrients which it may contain.

[1] Neilson-Jones, *Biological Aspects of Soil Fertility*. (See footnote [1], p. 79.)
[2] Rayner, *Mycorrhiza*.

As already explained, this elucidation by Rayner of the *complete* nature of compost action, including its ability 'to alter profoundly the soil bionomics', was quite incidental; her compost treatments, conceived for a special purpose, having no relevance to 'compost experiments' in the usual sense. It is for this very reason that it required something in the nature of an inspiration on Howard's part to see a connection between the forestry experiments at Wareham, and certain results being obtained in India on tropical crops.

This is how he himself describes this discovery:

'How does humus affect the crop generally and how does a factor like this increase resistance to disease? The large-scale trials of the Indore process now being carried out on coffee, rubber, cacao, and other crops in the tropics have furnished some interesting information on these questions.

'In a number of cases, in tea and rubber in particular, very striking results followed closely on *one* dressing of compost applied at the rate of five tons to the acre. There was a marked improvement in growth and also in the resistance to insect pests such as red spider, Tortrix and mosquito blight (Helopeltis). Two applications of compost have also transformed a derelict tea garden into something above the average of the locality.

'In a recent tour of tea estates in India and Ceylon I have seen these results for myself, and have discussed matters on the spot with the men who have obtained them.

'When these cases were first brought to my notice towards the end of 1936 and during 1937, I found considerable difficulty in understanding them. If humus acts as an indirect manure by (a) recreating the crumb structure and so improving the tilth, and (b) by furnishing the soil population with food from the use of which the soil solution eventually becomes enriched to the advantage of the crop, such factors would take time and we should expect the results, if any, to be slow. The improvement following humus was the reverse of slow—it was immediate and spectacular. Some other factor besides soil fertility appeared, therefore, to be at work.

'After much thought it occurred to me that the explanation would be found in the active root system of tea and rubber, and that the remarkable results recently obtained by Dr. M. C. Rayner on mycorrhiza in relation to forestry at Wareham in Dorset would apply to tropical crops.

'The simplest and most obvious explanation of the sudden improvement after one application of compost is the well-known effect of humus in stimulating the mycorrhiza which are known to occur in the absorbing roots of tea, and which in all probability are to be found in rubber, coffee, and other cultivated plants in the tropics.'

Howard's realization that mycorrhizal association might be a key factor in the proper nutrition of a very much larger range of plants than had hitherto been suspected, led to the careful examination of the roots of a large number of crop plants of both tropical and temperate climates. This revealed that a very large number indeed are mycorrhiza formers. The examinations were carried out by Dr. Rayner and her assistant Dr. Levisohn. A surprising parallel to the Wareham experience resulted.

In all the specimens examined vigorous healthy growth was invariably accompanied by extensive fungus infection of the roots of the plants concerned. This was defective or absent when the plants were poor and unhealthy. Crops grown with compost, or ample quantities of farmyard manure always showed maximum mycorrhizal development, in marked contrast to those grown with artificials.

In a circular letter to correspondents dated 15 July 1938, Howard lists some of the crops examined with the following notes on their reaction to different soil treatments.

'(a) *Cotton.* The Indian cotton crop provides an interesting pointer on the relation between a fertile soil and disease resistance and quality. In Central and South India thousands of examples of intensive cotton growing exist in the zones of highly manured land round the villages. Here cotton always does well and there is practically no disease: the yield and quality of the fibre are satisfactory. Surrounding these areas are stretches of extensively farmed land low in organic matter where the yield is uncertain, the quality average and the damage done by various pests often considerable.

'Many attempts have been made to increase the crop by means of artificial manures but the results I have seen have never been quite convincing: the benefits following chemical manures are not the same as those conferred by a fertile soil. If cotton is a mycorrhiza-former these differences would be explained. The mycorrhizal relationship would provide channels of sustenance between a soil rich in organic matter and the plant. The first of the appended reprints shows that cotton, like tea, is a mycorrhiza-former. In this note the bearing of this fact on production and future research is explained: the investigations of tomorrow in cotton will have to start from a new base line—soil fertility. Very much the same considerations apply to future research on tea.

'(b) *Grapes.* During a recent tour in the Midi some attention was given to the cultivation of the vine, a crop I studied for eight years on the western frontier of India. In the south of France grapes are raised very largely by means of artificials: the many diseases are combated by poison sprays.

'In Baluchistan, on the other hand, the vine is always manured with farmyard manure: artificials are not used: the crops have no need of fungicides and insecticides because disease is practically non-existent.

'On one occasion only near the village of Jouques, Bouches du Rhône, did I come across vines which at all resembled the vigour and health of those to be seen on the western frontier.

'These were manured with pig manure and had never been treated with artificials: the yield was excellent and I was told the vineyard had an excellent reputation for quality.

'Specimens of surface roots were found to be heavily infected with mycorrhiza. This interesting bit of Nature's machinery is used to good effect by the tribesmen of Baluchistan; it is largely side-tracked by the cultivators of France.

'(c) *Hops.* Several of the merchants who handle the hop crop have assured me that the quality has deteriorated since artificials have come in and that really good-quality hops can only be obtained by means of farmyard manure. Unquestionably the hop is much more liable to disease than it was in 1903–5 when I worked on this crop. In the course of some large-scale trials of the Indore process at Bodiam in Sussex samples of the surface roots of this crop were found to be heavily infected with mycorrhiza.

'(d) *Sugar Cane.* As is well known, what amounts to a new sugar industry has arisen of late years in the plains of India as a result of the research work done at Shahjahanpur and Coimbatore—two Experiment Stations I helped to bring into existence.

'The splendid crops at Shahjahanpur were always raised on humus and were to all intents and purposes free from disease.

'This was my experience at Indore on the black soils where the Shahjahanpur methods were followed. Reports have recently reached me from India that where artificials are used the cane is being attacked by insect pests. If the sugar-cane is a mycorrhiza former, a simple explanation of these differences would be provided. The first set of cane roots obtained from central India show that sugar-cane is a mycorrhiza former. Further examples from Louisiana, Brazil and Natal are expected shortly.

'(e) *Bananas.* The new banana industry of the West Indies is threatened with serious disease. In India I cultivated this crop successfully for some years and I also saw a good deal of the indigenous plantations. In all cases cattle manure was used and no disease trouble was ever brought to my notice. If the banana is a mycorrhiza former these results are understandable.

'The first set of banana roots I have just obtained from a rich garden soil in India shows abundant mycorrhizal infection.

'(f) *Strawberry.* The strawberry area near Southampton is rapidly contracting on account of poor soil conditions followed by disease.

'The small Experiment Station maintained by the Hampshire County Council at Botley is little more than a collection of pathological material. On the other hand commercial growers in this neighbourhood who use humus are getting satisfactory crops.

'Similar results are to be seen in South Lincolnshire and in other parts of England. Humus prepared from vegetable *and* animal wastes or farmyard manure produces healthy crops of good quality.

'The long continued use of artificials is followed by disease and indifferent fruit. The strawberry is a mycorrhiza former: the type of association being similar to that in tea. Subscribers interested in tea who grow strawberries can very easily compare the effect of humus and artificials on the health of the plants and on the quality of the fruit.

'These widely different examples—cotton, grapes, hops, sugarcane, bananas and strawberries—tell the same story and afford interesting confirmation of the views in the report on my recent tour of tea estates in India and Ceylon. Nature has gone to the trouble of creating a link between humus in the soil and the plant. It behoves us to make the fullest use of this provision.'

Rayner summarizes her observations on one of these examinations—that of cotton—as follows:

'The cotton plant is a regular mycorrhiza former, infection taking place under favourable conditions at an early stage. The incidence of infection is closely correlated with the nature of the rooting medium and in the same soil varies markedly with different manurial treatments.

'Differential behaviour on the part of the endophyte was particularly well-marked in respect to applications of inorganic as compared with organic manures, and a response was also apparent following the use of different forms of organic manures.

'That these differences should be observable is in itself significant, and the relation of the incidence and character of mycorrhizal infection with the growth and vigour of the hosts under crop conditions is obviously worthy of close attention and study.

'The observations are in general agreement with those made on *Citrus* in southern California some years ago and are believed to justify similar conclusions.

'It is not proposed to discuss here the character of the symbiotic relationship in the type of mycorrhizal association represented in cotton. It is evident that it differs fundamentally from that in ordinary parasitic attack and that there exists in the arbuscular-sporangiolar and vesicular apparatus a mechanism capable of functioning for nutritive exchange.[1] Coincidence of vigorous growth of the host with maximum infection may be observed in cotton as in other plants showing the same type of association, and this together with the histological evidence available supports the view that there is a substantial balance in favour of the vascular partner. Whether such exchange follows the lines elaborated by Frank and Stahl; whether it is concerned with major nutrients or takes a more subtle form such as that indicated in Butler's suggestion that "The vesicles and the arbuscular-sporangiolar apparatus are . . . mainly concerned with the accumulation of fatty material (possibly with accessory nutrients[2]) and its transference to the higher plant," is, in my view, of secondary importance at this stage of the inquiry.

'The immediate point of interest incidental to the cultivation of cotton or other mycorrhiza formers is to confirm the view here put forward, that the fluctuations in mycorrhizal behaviour that undoubtedly occur are causally related with changes in the root environment brought about by manuring and other soil treatments and the responses of the host plants thereto.

'My interest in the matter has been stimulated by the opportunity recently provided to examine root material of a large number of species grown as crop plants both under tropical conditions and in the British Isles, among others, tea, coffee, sugar-cane, oil palm, coconut, cocoa, hops, vine. The conclusions now offered have been confirmed and strengthened by these observations. In all cases for which comparative material was available—including a majority of the samples—that collected from plants recorded as specially healthy and vigorous showed a maximum development of mycorrhizas with structural features believed to indicate a balanced relationship.

'It is not hereby suggested that in all cases of regular mycorrhizal

[1] See Tech. Ref. No. 2, p. 89.
[2] See ibid.

association the habit is obligate or that some degree of mycotrophy is necessary for complete nutrition. The view held is rather that for species growing in nature the mycorrhizal condition in a healthy plant represents one of physiological equilibrium. Mycorrhizal association, often ignored or treated as an accidental and casual phenomenon, is a manifestation of biological soil activity, and as such cannot be profitably overlooked in the study of soil factors likely to promote healthy growth and maximum resistance to disease. . . .'

As already stated authorities are not agreed on the exact function of mycorrhizal association. Howard's view is that the root cells, acting very much like the stomach of an animal, obtain in the process of digesting the invading fungus, a protein food rich in nitrogen and phosphorus, and that:

'The mycorrhiza appears to be the machinery provided by Nature for the fungi living on humus in the soil to transmit direct to the active area of the roots the contents of their own cells. Whether this is the only means by which such things as accessory growth substances can safely pass from humus to plant, or whether the fungi provide essential materials for their manufacture in the plant itself, has yet to be determined with certainty. Some such explanation of what is taking place seems exceedingly probable. If the accessory growth substances contributed by humus were to pass from the soil organic matter into the pore spaces of the soil they would have to run the gauntlet of the intense oxidation processes going on in the water films which line these pores. In this passage any substance of organic origin would be almost certain to be seized upon by the soil population for food and oxidized to simple substances, such as the plant ordinarily takes in by the root hairs. If, as seems almost certain, freshly prepared humus (obtained from animal and vegetable wastes) does contain growth-promoting substances (roughly corresponding to the vitamins in food), it would be necessary to get these into the plant undamaged and with the least possible delay.

'The mycorrhizal association in the roots, by which a rapid and protected passage for such substances is provided, seems to be one of Nature's ways of helping the plant to resist disease.'

There is no direct proof that this theory is correct, but it is an interesting hypothesis. It is now recognized that the inclusion of a proportion of fresh raw food in the diet of man and beast is the surest way of supplying those vitamins essential to the avoidance of malnutrition and the deficiency diseases caused thereby. Will this rule prove to apply also to plants? If so, the fact that disease should follow the under-nourishment which would result from the omission of this living element from their diet, would be a perfectly logical sequence of events.

There is, however, a weak point in this theory, for it fails to explain the response to compost treatment of non-mycorrhiza formers.[1]

It seems more likely that it is the action of soil fungi in general, which, in some way—as yet unexplained—releases growth-promoting substances, and that mycorrhizal association is only one factor in the case, though of prime importance to mycorrhiza formers. It seems clear that this is the view which Rayner at present favours. Discussing the observed fact that introduction of active mycorrhizal material into the soil at Wareham was followed by an impetus to short root formation, this root growth *preceding* infection by the fungus mycelium—she states:

'Of the exact nature of the stimulus to root production nothing is known; the term "growth-promoting substance" is used in a quite general sense. Nor is it known whether the capacity to bring about such stimulus to root production is limited to those fungal species capable of forming mycorrhizal associations. To the writers it appears improbable that this should be the case. . . .

'The view reached is that mycorrhizal association is a highly specialized aspect of the ecological balance normally maintained between the higher plants and their fungus competitors. The study of mycorrhizal association as an isolated phenomenon may obscure its real significance as an edaphic factor, a liaison aspect of soil bionomics linking the microbiological activities concerned in the maintenance of soil fertility with the nutrition of vascular plants. The specific bacterial and fungal associations exemplified in nodules and mycorrhizas are special cases of a complex ecological system embracing members of the surface vegetation and those of the soil microflora.'

The leguminous crops which form nodules in association with the bacteria referred to, and thereby are enabled to assimilate atmospheric nitrogen, may prove to be a fruitful field for further research into this question. All the members of this family that have been examined are also mycorrhiza formers, but with crop plants of this family, such as pulses and clovers, nodules and mycorrhizas *have never yet been noted on the same rootlets.* These have either formed nodules only, or mycorrhizas only, maximum development of the latter being associated with those plants showing also maximum vigour.

There seems little doubt that the health and quality of this group

[[1] See Part II.]

of plants suffer when mycorrhizal association is absent. Howard, writing on this point, states:

'A long experience of the cultivation of leguminous plants in India has completely shattered my belief in the idea that these crops can be grown successfully without organic matter, and that the nitrogen fixation in the nodules is the complete story as far as the supply of combined nitrogen is concerned. Farmyard manure or compost, as already stated, is essential for keeping these crops healthy and for making them form seed in the Indian Monsoon. Organic matter always stimulates both root and nodular development. . . .

'There is a further point of some interest in this matter. When plants like French beans are grown on poor soil by means of the nodules only, or by means of artificial manures, the produce is tasteless and of poor quality. For real taste and quality in the produce it is necessary to use humus (made from both vegetable *and* animal wastes) or farmyard manure. A supply of combined nitrogen appears therefore to reach the plant by way of nodules and root hairs; the materials which are needed for quality appear to be absorbed by the mycorrhiza.

'The leguminous plant therefore promises to be a very valuable instrument in separating out the various factors concerned in this question. Will, as seems to be the case, quality and disease resistance only be obtained when the mycorrhiza mechanism functions? Will disease resistance and quality turn out to be the same thing?'[1]

It is of vital importance that a correct answer to this question should be found, but without waiting for the research which will assuredly one day provide it, one can go so far as to say, I think, that the evidence already submitted here, justifies the conclusion that mycorrhizal association is a habit that can no longer safely be ignored in the study of the nutrition of crop plants. Our first question has thus been answered in the affirmative.

The next three points on the list can be taken together. They all concern the toxic condition of Wareham soil. At first sight it would appear extremely improbable, if not impossible, that any parallel to this condition should be found in cultivated soils. Two things, however, are noteworthy, first, that while Wareham is an extreme case of its kind, its condition is one of degree rather than of essential peculiarity[2] and second, that even in cultivated soils certain soil

[1] Howard, 'Insects and Fungi in Agriculture', *Empire Cotton Growing Review*, 1938. [See also Part II.]

[2] Neilson-Jones, *Biological Aspects of Soil Fertility*. (See footnote, p. 79.)

treatments have definitely been shown to depress mycorrhizal activity, and therefore it can be assumed that inhibition of fungal activity generally, *can* exist in such soils.

Our next question then, can be formulated thus: Does any evidence exist that where fungal activity is inhibited in cultivated soils a toxin may be operating of a kind comparable to that present in Wareham soil?

There can obviously be no great body of direct evidence on this point until many different soils have been subjected to the same type of intensive investigation as that carried out by Professor Neilson-Jones on Wareham soil, but a close study of his most interesting report of this investigation, notably in regard to some of the methods employed to increase and decrease toxicity, and a comparison of some of his results with certain causes and effects well known in farming practice, give rise to a not inconsiderable weight of circumstantial evidence. The Neilson-Jones experiments prove that the toxin indirectly inimical to the growth of higher plants in Wareham soil, is a by-product of some soil organism which operates directly by inhibiting fungal growth.

In the course of the investigation strong evidence was forthcoming that the organism concerned is anaerobic in habit. Waterlogging for example greatly increased the toxicity; partial sterilization by steaming for short periods, or by air drying, had the effect of inactivating the organism, but toxicity was regained when waterlogged conditions were restored. It required complete sterilization by steaming for an hour or more, baking under pressure, or soaking for long periods in alcohol to kill the organism. Soil so treated in the laboratory did not regain toxicity spontaneously under waterlogged conditions but required re-inoculation with untreated soil before doing so.

The identity of the organism responsible has not yet been definitely established, but the group of bacteria capable of reducing sulphates was suspected. This group was found to be present in Wareham soil in unusually large numbers. In the process of reducing sulphates these bacteria produce the poisonous gas sulphuretted hydrogen (H_2S). They can produce it either from sulphates applied as chemical salts, or from imperfectly decomposed organic matter, of which, as will presently be shown, there is also a high proportion in Wareham soil.

Although the presence of H_2S in Wareham soil has not been proved by direct chemical tests, all the observations and experiments made with soil samples in the laboratory are in conformity with the view that this gas may be responsible, at least in part, for the arrest of fungal growth. An account of one such experiment will suffice here.

'A series of cultures of once-steamed soil was set up in flasks or Petri dishes of which half received an addition of 0·1 per cent NaCl (sodium chloride) solution, the remainder a corresponding addition of Na_2SO_4 (sodium sulphate) solution, each culture receiving sufficient to produce a waterlogged condition. A period of twelve days was allowed to elapse during which stimulation of the activity of H_2S-producing bacteria might take place, after which all cultures were exposed to air infection. A month later, all flasks that had received NaCl solution showed a profuse growth of *Mucor* spp. and *Penicillium* spp., whilst those that had received Na_2SO_4 solution showed either no mycelial growth or very sparse growth of *Penicillium* spp., any fungal development that had appeared in the earlier stages having retrogressed.

'The increased toxicity of soil to which sulphate is added is explicable if the toxicity is related in any way with a slow and continuous evolution of traces of H_2S derived from the activity of bacteria capable of reducing sulphates.'

Neilson-Jones adds that: 'This interpretation is purely speculative in that it is based on an isolated observation'; but he then proceeds to describe further tests and observations, all corroborating the one just described, in the course of which it was shown that the mycelium of airborne fungi 'is highly sensitive to H_2S, the smallest trace being sufficient to check growth.'

Many soils contain sulphur-reducing bacteria, particularly clay soils. It is well known to cultivators of clay soils that good crops usually follow a well-made bare fallow. Bare fallowing does not mean simply leaving a field derelict, as many laymen seem to think; it is, on the contrary, an expensive preparation for a subsequent crop, and consists of several ploughings, and constant surface cultivation throughout the summer. The effect of this, in addition to killing weeds, is to thoroughly air dry the top six or more inches of soil.

In his investigation on Wareham soil Neilson-Jones found that 'Air drying has an effect on soil somewhat similar to steaming: on remoistening the toxic reaction is found to have disappeared and mycelium grows as readily though with less vigour than on steamed soil. In fact, complete air drying may be regarded as a mild form of partial sterilization, the capacity of the soil for supporting a vigorous superficial fungal growth subsequently being due as with steamed soil, to the removal of the toxic reaction of the soil and the provision of an additional source of food—in this case presumably from the organisms killed by desiccation.'

There is an old saying that a bare fallow is as good as a coat of muck.[1] It appears that this may be literally true in that both have the power to discourage organisms capable of producing a toxin inimical to fungal growth, the difference being that the effects of one are temporary, while the other produces conditions which tend to be self-propagating and therefore permanent, or semi-permanent.

Here then, in two methods proved to be effective in removing the toxic reaction from Wareham soil, namely compost treatment and air drying, we have a close parallel in ordinary farm practice. Farmyard manure, and the bare fallow. What of the conditions shown to increase toxicity?

The deliberate waterlogging in the laboratory tests also has its counterpart on the land.

That it is impossible in ordinary farm practice to grow good crops on waterlogged land is a truism the reason for which is recognized as being a question of aeration. Where the pore spaces in the soil are filled with water, air is excluded. 'Absorption by roots, being a process dependent on the action of living cells, is directly affected adversely by absence of oxygen,' but Neilson-Jones points out that 'a survey of the area at Wareham Forest made it clear that the inhibition of growth observed there was not due solely to such direct action. The work now recorded shows that factors of many kinds are concerned, and provides evidence that plant growth may be affected adversely or the reverse through causes inherent in the soil bionomics, members of the soil population showing great sensitiveness to aeration and to the nature of additions to the substrate whether derived from their own metabolic activities or from external sources.'

Once more we have to question whether this state of affairs is peculiar to Wareham soil. Here I think I can furnish some evidence from my own observation. I know a farm in Scotland, one lying in a notably fertile area, the land of which has been associated for generations with heavy crops, and noted for the excellence of its sheep and cattle. A few years ago this farm became part of a much larger unit the whole of which was turned over to mechanized grain crop production. The livestock disappeared, and in their place was seen the artificial manure distributor, and the harvester-combine. Temporary leys

[1] Bare fallowing is, of course, not suitable to all climates. Soil aeration is frequently best achieved by use of the wheel subsoiler and by growing deep-rooting crops.

have recently been reintroduced on this farm, and a considerable head of livestock is once more being kept. But towards the end of its stockless period, although the land was still yielding heavy crops a definite increase took place in certain weeds usually associated with waterlogged soils, or soils with a pan. Now on this land the appearance of these weeds was certainly not due either to bad cultivation or to waterlogging, nevertheless, a condition developed which pointed to faulty aeration. What is the factor, or factors, responsible? Possibly they include a reduction in the earthworm population. This usually occurs where artificials have been substituted for farmyard manure, and earthworms are great soil aeraters.[1] Conditions inimical to earthworms are equally unfavourable to fungal activity. Both worms and fungi appear to be highly sensitive to certain chemical salts[1] notably sulphate of ammonia. It seems a plausible hypothesis therefore, that the appearance of conditions suggesting faulty soil aeration may be connected with poisonous products resulting from the interaction between sulphate-reducing bacteria and ammonium sulphate applied as fertilizer.

In stockless farming I am told that the bare fallow is an absolute necessity. (On stock farms the well-mucked root crop, or the green crop folded by sheep usually takes its place.) In view of the evidence already presented in this chapter this is at least suggestive.

In the case of another mechanized farm, this time in East Anglia, the owner a few years ago boasted that there was not a single animal on the place, and that he had been growing heavy crops of wheat and barley continuously, with the help only of large dressings of artificial fertilizers. A year or two after this, the expert who advised him as to his fertilizers told me himself that the ravages of 'Take-all' in the wheat on this farm had reached such disastrous proportions, that he had advised the ploughing up of all corn land and putting the entire acreage into sugar-beet.

There is no question but that this disease is on the increase; the chemists are trying to find a seed dressing or spray that will stop it, but this is on a par with looking for an anti-toxin for a deficiency disease. Even if it succeeds, you are but curing a symptom. The two examples I have just given are merely straws in the wind, but the evidence is considerably strengthened by an incident that occurred on my own farm. Howard's view is that 'Take-all' in wheat is a deficiency disease, and that the principal primary cause is lack of

[1] See Tech. Ref., p. 112.

aeration, this interfering with normal mycorrhizal association. The following personal experience supports this view to a startling degree. I have an 18-acre field, which was, for many years, badly in need of draining; in winter it was invariably waterlogged in patches. The soil in question is medium clay loam on a clay subsoil with pockets or drifts of sand in it. In the course of time drains passing through these sand gaults collapse and a kind of pond is formed in the area. There were several of these places in the field varying in size from a few square yards to as much as half an acre and even more. I worked the field for some years before being in a position to drain it, and came to know all these patches well, for whenever the field was sown with winter grain such as wheat, the crop never survived on these patches, which produced only a kind of water-grass. The outline of these places was irregular in shape, but well-defined, and year after year, exactly the same outline was invariably visible in the growth of the water-grass, and the absence of the crop. One year, shortly after I had first met Howard, and studied his work, we had an exceedingly dry autumn, winter and spring. The field was under wheat, and as that year there was no waterlogging because there was no rain to speak of, the crop germinated and grew evenly over the whole field. At the flowering stage there was no observable difference between the wheat on the 'wet' places and that on the rest of the field, but by harvest time the wheat growing in the patches normally waterlogged, was all blighted with 'Take-all'. The area of the diseased wheat followed exactly the outlines of the badly drained patches. All the rest of the wheat was perfectly sound. So much for the one-time theory that the disease is air-borne.

Now the patches were not waterlogged that year, but by-products of past years of anaerobic conditions were almost certainly still present, and it is safe to say that although the plant roots must have obtained sufficient oxygen in that season, fungal growth, even if inhibitory factors were not present, would not have re-established itself in the time.

Neilson-Jones makes this observation in connection with his investigation of Wareham soil:

'Chief among the initial causes of inhibition of mycelial activity in natural soils is probably low oxygen pressure which depresses fungal activity directly and may do so indirectly from the nature of the by-products of anaerobic metabolism. Deleterious by-products in the soil may be slow to disappear, even when the causes responsible for their accumulation, such as lack of oxygen, are no longer operating.'

My experience suggests that this applies also to cultivated soils.

These are some of the examples of circumstantial evidence which bear on points two, three and four, and I think it must be admitted that they suggest, if no more, that the lessons of the Wareham researchers, in connection with soil toxicity, may well be applicable to a much wider field.

The evidence concerning the fifth and sixth points on the list can also be considered jointly. These present the question as to whether, in the case of ordinary crop plants, compost acts directly on the roots by supplying plant nutrients, or, as at Wareham, indirectly by stimulating the activity of soil fungi.

On the relationship between fertility and soil fungi in general, Neilson-Jones has this to say:

'Study of the causes of soil infertility in Wareham Forest justifies the generalization that in organic soils of the acid type, possibly in all soils, maintenance of the activity of fungi, whether those responsible for the breakdown of organic detritus or those specialized forms concerned in the production of mycorrhizas, is of critical importance. It is not suggested that organisms belonging to other groups are not vitally concerned, but it appears that fungi form an essential link in a mechanism whereby organic detritus is incorporated into humus of a fertile soil and that therefore their activity can be used as an index of fertility; infertility may result either from absence of appropriate fungi or from the presence of factors hindering their activity.'

Lack of aeration is one such factor operating not only directly but also indirectly by favouring the activity of such anaerobic organisms as sulphur-reducing bacteria, which, as mentioned above, can produce H_2S or other volatile sulphur compounds either from sulphates applied to the soil as chemicals, or from organic matter when normal decomposition is arrested. Such activity forms a vicious circle, since the H_2S produced by the bacteria inhibits the activity, among others, of those fungi responsible for the decomposition of organic matter, and the resulting arrest of normal decomposition further encourages the production of H_2S.

One effect of compost is to break this vicious circle, and the enormously increased rate of cellulose decomposition in soil to which compost has been added is one of the proofs of its action through the medium of soil fungi.

In an experiment designed to measure this factor cotton-wool pads of known weight were buried for four months in untreated Wareham soil, in ordinary woodland soil, and in Wareham soil plus C 5 compost.

At the end of the period what was left of these pads was dried and reweighed. Representative results based on many repetitions of the experiment showed that in untreated Wareham soil only 10 per cent of this cellulose had been decomposed, in the woodland soil the figure was 33·6 per cent, but in Wareham soil plus compost the percentage of decomposition was over 91 per cent. This experiment indicates a very simple way of testing for fungal activity, and the effects upon it of different soil treatments. It could easily be undertaken on cultivated soils as a regular routine, in conjunction with ordinary field analysis tests for determining soil requirements, and as a measure of fertility. I know of only one place where any attempt to carry out such a test on agricultural land has been made. These tests have not been going on long enough yet for the results to have much significance, but so far the indications are that the highest rate of cellulose decomposition would be found in fields that had received compost, and the lowest in those that had received artificials.

This matter of cellulose breakdown is of vital importance to the whole question of the nutrition of living cells, for it has been shown by experiment that raw cellulose can be as injurious to plants[1] as it is indigestible to animals, and in both the soil and the digestive tract of herbivora, cellulose decomposition is largely performed by fungi. (See Chapter 6.)

Such soil treatments, therefore, as ploughing in straw or green crops can only prove successful if conditions for the subsequent breakdown of the cellulose are present. It is useless to apply organic matter in this form while at the same time inhibiting, by the use of certain artificial fertilizers, the fungi capable of converting that organic matter into humus.

The increasing use of harvester-combines even on moderately sized farms makes this a question of very particular importance, for the problem of such farms is the disposal of the straw left on the field by the harvesters. Burning is wasteful, and ploughing it in not only presents mechanical difficulties, but is not found particularly satisfactory in its results on the subsequent crop—sometimes definitely

[1] Of great significance also are the changes in resistance to attack by soil fungi that accompany alteration in the soil environment. In Wareham soil a severe attack on the roots of pine seedlings by *M. atrovirens*, here normally innocuous, can be induced by the addition of 5 per cent of pure cellulose to the soil of pot-cultures. In another heath soil of similar type, *Rhizoctonia silvestris* behaves in like manner, producing pseudomycorrhizas encrusted with massive sclerotia on addition of 5 per cent pure cellulose. In both cases the health of the seedlings suffers severely. (Rayner.)

the reverse.[1] This adverse factor is not difficult to understand when it is remembered that farms using such implements are usually those on which large quantities of chemical fertilizers are also used, and where nitrogen is given largely in the form of sulphate of ammonia. Sulphates, as we have seen, encourage toxin-producing organisms which inhibit the fungi needed to break down this straw. There are various ways of overcoming the difficulty, one of the most effective being to under-sow the corn with some leguminous crop which is allowed to grow through the scattered straw and then fed off by livestock, but on the stockless farm the 'solution' sometimes adopted is to sweep the straw left by the combine into a heap and to compost it according to the recognized formula of 1 lb. of nitrogen to 100 lb. of dry matter; but the nitrogen generally used for this purpose is sulphate of ammonia! True, this has the effect of breaking down the cellulose and producing a finished product of well-rotted material, not unlike other compost in appearance, but, as the Wareham experiments clearly show, the value of a compost depends upon its organic constituents, and not upon its appearance or its chemical analysis. Failure to recognize this is responsible I think for some of the discrepancy of view regarding the value of compost. The source of the nitrogen used in composting appears to be of considerable importance to the value of the final product. Composts made with chemical activators may increase yield in crops, just as chemicals alone also do—for a time—but the growth of healthy crops showing *maximum disease resistance* has, so far as I can discover, only been obtained when the activating agent is of animal origin. The one exception seems to be in the case of compost the raw materials of which consist of mixed garden refuse including leaves and young weeds in full growth. Such compost can be activated by certain herbal extracts such as those developed by Miss M. E. Bruce with results in the subsequent crops apparently in all respects equal to that of animal-activated compost. Such raw materials are, of course, extremely rich in a wide range of trace elements and the resulting compost is rapidly invaded by an immense earthworm population which automatically introduces an animal element. For her work with conifers Rayner found that dried blood produced far better results than any other activator used. The best activating agent for Indore compost for ordinary crops in this country is, in my experience, fresh farmyard manure in which the urine is preserved; this last is very important. I have, however, obtained quite satisfactory results

[[1] See footnote [1] to p. 14.]

by composting with both raw and coagulated blood, and also with sewage sludge, alone and in mixture with muck or blood. The point is that for maximum results in the quality of the crop subsequently grown with it, compost must include some product of animal origin. People who advocate sulphate of ammonia or 'Adco', or other inorganic products for breaking down cellulose in the compost heap, are still suffering from the illusion that humus nourishes the plant direct. The Wareham experiments clearly show that this is not the case. Rayner's findings in this respect were fully endorsed by the independent investigation cited in this chapter.

'That the fundamental effect of the compost is to alter profoundly the soil bionomics is proved by the enormously accelerated rate of cellulose breakdown and the stimulation of the fungus flora in general in soil that has been treated with compost. That the basis for these changes lies in the organic constituents of the compost rather than in the inorganic nutrients it contains is evidenced by the efficacy of leached compost, by the differential effects of composts differing only in respect to organic constitution, by the dissimilar effects wrought by the same composts on different organic soils, and by direct experiment with additions of salts of nitrogen, potash and phosphoric acid equivalent to those contained in a given compost (Rayner, 1939). It is suggestive that the growth-promoting action of compost on fertile garden soils is often negligible or comparatively small; here presumably conditions are already favourable to fungal activity so that additions of compost can do little to improve fertility by modification of the micro-flora.'[1]

This supports the view that the effectiveness of compost in cultivated soils depends, just as it does at Wareham, not on its value as a direct plant food, but on its capacity to stimulate the activity of soil fungi. This explains why adherence to fundamental principles in the making of compost heaps is so important, and also why chemical activators fail to produce the degree of disease resistance in plants that result from the use of such composts as those advocated by Rayner and Howard, and also why negative results are reported by users of the putrefactive heaps misnamed compost. Chemical activators will quickly rot vegetable rubbish, but there is much more to the manufacture of humus than this. It involves the creation of what is at once a habitat and a food for soil organisms, but more than this, it must be so fashioned as to favour the activity of the beneficial species and discourage others. There are many species involved beside those

[1] Neilson-Jones, *Biological Aspects of Soil Fertility.*

responsible for cellulose decomposition, the mycorrhizal fungi forming at least an equally important group.

These various fungi feed on the humus in the soil, and it is the product of their metabolism which is of such vital importance to the complete and balanced nourishment of the plant.

Once this is understood, the necessity for the inclusion of animal residues in the compost heap at once becomes clear, for fungi require a mixed diet of organic matter. Not only are they unable to take their nourishment exclusively from inorganic chemicals, but they are not even purely vegetarian.

'In 1894, so noted a botanist as Hooker, speaking of the fungi, observed: "These plants seem to invert the order of Nature and to draw their nutriment, in part at least, from the animal kingdom, which it is held to be the function of the vegetable kingdom to sustain." '[1]

Thus it will be seen, that once you accept the view that soil cannot be fertile when soil fungi are inhibited, then Howard's view on the value of compost and the harm done by many chemical fertilizers and poison sprays, is shown not to be a fad, as some people seem to think, but a perfectly logical conclusion based on practical experience, and supported by scientific evidence.

I sum up this chapter with a final quotation from the paper already cited.

'The prescription for compost treatment advocated by Rayner to correct a special case of infertility in a natural soil clearly envisaged removal of unfavourable by-products, with alteration of the soil processes responsible for producing them, in addition to promotion of mycorrhiza formation. The remarkable and persistent improvement in tree growth that followed from this treatment brings the experimental results at Wareham into line with records recently provided describing the action on crop growth of composts manufactured by the Indore process when applied to cultivated soils in many parts of the world (Howard, 1940). The great emphasis laid by Howard on the significance of mycorrhizal activity is a welcome and long-overdue recognition from the practical side of the important part played by this habit in crop nutrition; the indifference displayed to this aspect of soil research by soil specialists is difficult to understand. But the mycorrhizal habit, although widespread is not universal in vascular plants; the soil possesses fertility in its own right as reflected in the growth of non-mycorrhizal as well as mycorrhizal plants. *It is the free*

[1] Rayner, *Mycorrhiza.*

action of fungi that is regarded as essential for the development of soil fertility;[1] that some of the species concerned also form mycorrhizal associations with certain of the higher plants may be to the advantage of the latter, but this is additional to and different from the generalized service that fungi, mycorrhizal or not, perform in the soil—it provides an alternative channel uniting mycelial activity in the soil with the nutritive processes in vascular plants.

'The analyses of a particular case recorded in the present researches shows very clearly some of the ways in which a condition of biological inertia in the soil, however produced, may bring about infertility for growth of the higher plants, and has yielded direct evidence of some of the ways by which additions of organic material operate in restoring fertility.

'It is realized that there are gaps in the argument as presented in this paper; it could hardly be otherwise in view of the complexities involved. Nevertheless, the evidence is so striking that disturbance of microbiological equilibrium in the soil by inhibition of fungal activity is a potent factor in bringing about infertility for growth of vascular plants, its direct relation with the results of forestry researches so impressive and the possibilities of wider application so evident, that publication at the present stage of the work appeared to be justified.'[2]

It seems that the equation in the 'Medical Testament' should be redrawn thus:

$$\left.\begin{array}{l}\text{Vegetable}\\ \text{Animal \&}\\ \text{Human}\\ \text{Waste}\end{array}\right\}\rightarrow\text{Fungus}^{3}\left\{\begin{array}{c}\rightarrow\text{Soil}\rightarrow\\ \\ \longrightarrow\end{array}\right\}\text{Plant}\rightarrow\text{Food}\left\{\begin{array}{c}\rightarrow\text{Animal}\rightarrow\\ \\ \longrightarrow\end{array}\right\}\text{Man}$$

Technical Reference

THE EARTHWORM

F. H. Billington, in his book *Compost*, mentions how as a result of the interest aroused by the three-volume book, *Our Friend the Earthworm* by Dr. G. S. Oliver of Texas, and the latter's enterprise in breeding earthworms to supply to farmers, and for other purposes, 'much barren land from which the earthworm had been banished by

[1] Italics mine (author).

[2] Neilson-Jones, *Biological Aspects of Soil Fertility*, now published as part of *Problems in Tree Nutrition*.

[3] The word *Fungus* is here used to denote various groups of fungi in association with other micro-organisms.

chemical manures and poison sprays, has been restored to fertility'. He gives the American *Nature Magazine* of January 1941 as his authority for this statement.

He then supplies the following facts which make such claims understandable:

'(a) Darwin found that in an average English soil, earthworms brought to the surface about ten tons of castings per acre per annum. These are now known to consist of neutral colloidal humus, the only form immediately available to plants. It is safe to assume that in a warmer climate, or with improved types of worms such as the 'Coolie', this figure might be much larger.

'(b) The estimated numbers of earthworms per acre in the soils of the Rothamsted Experimental Station, England, are stated to be, very roughly: (i) 0·5 million on unmanured land; (ii) 2·75 millions on farmyard manured land; (iii) 8·6 millions on grassland.

'(c) Earthworms render soil permeable to rain thus checking the tendency to erosion by rain and wind. Aeration and nitrification are also stimulated.

'(d) They mix organic matter thoroughly throughout the soil and prevent its undue accumulation in peat-like layers on the surface, e.g. old, matted, sour pastures.

'(e) Earthworms are excellent subsoilers; what they bring up of this soil is vastly different to the crude material resulting from subsoil tillage.

Soil and organic matter swallowed by the worms are thoroughly pulverized in their fowl-like gizzards, and mixed with digestive secretions.

'(f) Earthworms perform equally useful work in the compost heap—abundance of them is a sure sign that the process is going well, just as their eventual withdrawal, in a normal way, is an indication that the compost is ready for use.

'Earthworms are very sensitive to acidity or purely chemical salts, whereas such materials as the wastes of the onion family and natural sugary materials, encourage them.'

6

Indications

[A number of the examples given in this chapter, which were topical at the time, can no longer be seen, owing to the death or departure of the persons concerned. Current literature has, however, recorded plenty of up-to-date examples of a similar kind in all three categories (Plant–Animal–Man) able to replace and supplement those cited here. A word should perhaps be said about the paragraphs concerning foot-and-mouth disease on pp. 136 and 137. The form of the disease (and we now know several viruses are involved) is different in India from, and almost certainly less virulent than, that in England. In one of the worst recent epidemics here, a very good organic farmer had two farms in the centre of a badly infected district. One farm escaped, the other succumbed. In the latter case the infection was definitely traced to a mass dose carried by a human being.]

I

PLANTS

In the last two chapters I sought to explain the complex character of compost action, and to show that humus in the soil benefits the plants in three ways: mechanically, as a direct plant food, and by fundamentally modifying the soil bionomics. Of the three, this last, hitherto largely ignored, is probably the most important.

I have shown that the action of soil fungi plays a vital part in the nutrition of certain plant species, and have given evidence of strong indications that the health of all plants growing in soil is affected to a greater or less degree by the activity of the soil micro-flora in general, and that it is the natural habit of the great majority of our crop plants also to form direct association—probably symbiotic—with specific soil fungi. Keeping this aspect of plant nutrition in mind, we are now in a position to consider the value of further evidence bearing on the relationship between soil fertility and health.

In Chapter 1 I gave a summary of the far-reaching claims made by many advocates of humus farming. Let me remind you what these are. That if the fertility of the soil is built up with adequate supplies of humus, crops do not suffer from diseases and do not require poison sprays to control parasites; that animals fed on these plants develop a high degree of disease resistance, and that man, nurtured with such plants and animals can reach a standard of health, and a power of resisting disease and infection, from whatever cause, greatly in advance of anything ordinarily found in this country.

Some mention of the evidence on which these claims are based has already been made. It is now time to consider a wider field of such indications as exist in support of them.

Let us take plants first. The claim speaks of resistance to disease and parasites. We must consider the evidence under these two categories separately.

By plant disease, is usually meant those ailments which are caused by virus, bacteria, or special parasitic fungi. It is well known that a shortage of certain essential soil ingredients, besides causing derangement in the plant's metabolism, can also induce some of these disorders; in fact in some cases the particular soil deficiency can safely be deduced from the appearance in a crop of the disease in question. As an example, 'finger and toe' in turnips indicates a lack of lime in the soil, and very many diseases are now explained by lack of some essential trace element. Deficiency diseases, in other words, are not confined to the animal kingdom.

This being so, it does not require a great feat of credulity to accept the view that *if* any aspect of normal plant nutrition depends on fungal activity, then the disease-resisting capacity of any plant will be weakened by removal of this element from its diet. Particularly should we expect to find such lowered resistance resulting from inhibition of mycorrhizal fungal activity in those plant species which normally form mycorrhizal associations. Yet the condition of mycorrhizal activity is seldom taken into account by plant pathologists.

'No true picture of the soil as an environment in which the root system of vascular plants passes its life can be formed if there is omitted a biological component so frequently present and often so abundantly developed as the mycorrhizal system. For example, although there may be at present no information as to the interaction of this habit with the organisms, and soil factors, operative in root diseases, yet it would appear dangerous to assume none to exist. Nevertheless, instances are not wanting in recent researches in which the normal condition of the root tissnes in respect to mycorrhizal infection, even in well-known and admitted mycorrhiza formers, is completely ignored in pathological investigations.'[1]

In cases of disease due to parasitic fungi, there are good grounds for believing that the power of the host plant to resist attack may be closely bound up with mycorrhizal association. In discussing this point with regard to pines, Rayner states:

'Study of these soil inoculation experiments raises a matter of practical importance in relation to mycorrhizal associations in general. It is already clear that the damage inflicted by soil fungi such as the forms of *M. r. atrovirens* and "*Rhizoctonia*" now under discussion does not depend merely upon the propinquity of mycelium to the roots. Apart from direct effects of the soil environment it is conditioned by other factors such as the vigour of the root system as a whole, the presence of suitable mycorrhiza formers, and the capacity of the host for mycorrhizal association. In cases where root response is satisfactory in these respects, there appears to be more or less complete immunity from parasitic invasion.'

Elsewhere she writes:

'Discussing this matter in reference to certain plant diseases, especially those of sugar-cane, it has been pointed out recently that disturbance of the normal activity of the mycorrhizal fungus may lead to secondary invasion of roots by bacteria or parasitic fungi, e.g. species of *Marasmius* or *Rhizoctonia* (Constantin, 1924).'

My own experience recorded in the last chapter, concerning 'Take-all' in wheat on the badly drained field, is very probably a case of this kind. The interaction, if any, of the mycorrhizal fungus with the activity of the fungus responsible for 'Take-all' has not, so far as I know, been studied, but since wheat is a mycorrhiza former it is more than likely that some such connection exists.

In fact, once accept as a likely hypothesis that under normal conditions plants derive some part of their food (something possibly analogous to vitamins), from substances resulting from the activity of certain soil fungi (and in view of the evidence it is difficult not to accede so much), then the recorded cases of plant disease being cured or prevented by compost treatment are readily explicable. That many such cases do exist is incontestable. Examples have already been given in respect to tea, cotton, grapes, sugar-cane, and strawberry. This year Howard records the following further experience with strawberries:

'I grew in heavily composted soil a collection of Royal Sovereign strawberries, badly infected with a common virus disease, alongside

[1] Rayner, *Empire Cotton Growing Review.*

some healthy stock. This year, 1942, the strawberries raised from these two sets of plants were about the best I have ever tasted. I found no trace of the virus disease. Similar results have been obtained by several of my correspondents.'

Results of the use of compost are in fact now available from all over the world. Here are a few representative examples. Mr. R. Paton of the Morib Plantations Ltd. of Banting in the Federated Malay States, writes:

'We started to keep livestock on a fairly big scale in 1930 for the purpose of manuring our coconuts, and this was done in conjunction with composting of husks, fronds, etc. in trenches two feet deep along the centre of each row. . . . Our average yield per acre was below nine piculs of copra, and the palms were then beyond the age at which one would expect any appreciable response in yield. Nevertheless, they have yielded over fourteen piculs per acre average for each of the past five years, and look like doing even better. Fine results have been obtained also in our rubber areas, particularly in young replantings, where the growth is all that could be desired, and not one ounce of artificial fertilizer has been used.'[1]

A good example from Europe was to be seen before the war on Dr. Pfeiffer's farm, near Flushing in Holland. Neither his greenhouse nor his outdoor crops received any artificial fertilizers. He succeeded in producing heavy crops of a flavour and keeping quality far in excess of those in the surrounding neighbourhood. Moreover, he never needed to use poison sprays because he was not troubled with insect or fungus diseases. Dr. Pfeiffer follows the biodynamic method of agriculture advocated by the late Dr. Rudolf Steiner. It includes the production of high-quality compost, fortified by various herbal and other organic preparations.[2] Dr. Pfeiffer is now engaged in important research work in the United States of America.

Africa has provided many indications in recent years. In a foreword which Lord Bledisloe contributed to the pamphlet on this subject (referred to in Chapter 4) he records 'the amazing justification' of the Indore process of humus manufacture which met his eye 'in every direction' when he visited south Central Africa in 1938 as Chairman of the Rhodesian Nyasaland Royal Commission. 'There I found,' he says, 'on the one hand, vast areas of desiccated and

[1] *Compost News Letter No. 1.* (A four-monthly periodical published by the County Palatine of Chester Local Medical and Panel Committee.) Later incorporated into *Soil and Health.* [No longer published.]

[2] E. Pfeiffer, *Biodynamic Farming and Gardening*, published in the UK as *Soil Fertility, Renewal and Preservation*, Faber & Faber, London, 1947.

exhausted land, utterly destitute of humus, which provided sustenance for neither man nor beast and where disease was rampant among the native tribes, and on the other hand, in striking contrast, bumper crops of tea, coffee, cotton, tobacco and maize being harvested by natives who were relatively healthy, the secret being the use of compost, made (generally by the Indore system) out of decayed vegetation associated with animal and human residues.

'The fact that the discoveries of McCarrison indicated that immunity from degenerative human disease followed the ingestion of a fresh, well-balanced diet of unprocessed natural foods, and that those of Howard disclosed a marked similar resistance capacity, to both degenerative and infectious maladies, as a consequence of returning to the soil a sufficiency of carefully prepared waste products may be regarded as something other than a coincidence. In the latter case both crops and livestock showed this salutary immunity.'

From Rhodesia Captain Timson reports 'that the resistance of maize to the attack of witchweed, a flowering semi-parasitic plant, was raised to a point approaching immunity by manuring with Indore compost.'[1]

This result was confirmed by Captain Moubray of Chipoli who also records (1942) that in an exceptionally dry season (only 2 inches of rain fell between the end of January and middle of March) the pollen on the maize 'just dried up' nevertheless 'lands throughout the colony that were well composted stood up well and gave good results. The government', he states, 'are now appealing to everyone to make as much compost as possible.' Another 'outstanding fact' has been the seed production from composted sunn hemp as compared with that where no compost had been added. This seed crop he reports as being very bad throughout the colony, the average yield being only about one bag per acre, but 'notwithstanding bad rain conditions' his own composted crop yielded three bags an acre. He gives it as his considered opinion that it is not really profitable to grow sunn hemp for seed without the addition of compost.[2]

Coming nearer home, the effect of organic manuring on quality in vegetable growing can be seen on a large scale on Mr. Secrett's farm at Walton-on-Thames. 'Mr. Secrett uses practically no artificials and raises his produce on fermented stable manure. He stands at the head of his profession as regards quality.' (Howard.)

Lord Portsmouth has made an interesting contribution to this

[1] Howard, letter to *The Times*, 10 August 1939.
[2] *Compost News Letter No. 4*.

question in connection with keeping quality. Wheat straw is used for thatching many of the cottages on his Hampshire estate. By keeping careful records over a number of years he found that, although the crops in question were grown side by side on the same type of soil, thatch made with straw from wheat grown with humus lasted twice as long as that grown with artificials.

One of the most impressive examples of large-scale humus farming in this country is to be found at Surfleet near Spalding in Lincolnshire, on the Iceni Estate, owned and farmed by Captain R. G. M. Wilson. This particular farm, consisting of nearly 300 acres, is of special interest because of the interdependence of its market garden, and more normal farming departments, but in connection with the special subject matter of this chapter, I am concerned with a result which he has noted in connection with the use of compost[1] on glasshouse tomatoes. This case is of interest because it is an example of the beneficent effect of compost on a presumed non-mycorrhiza former. When he first started growing tomatoes on a large scale, Captain Wilson tells me he was much troubled by that uneven ripening of the fruit which is such a frequent cause of loss to commercial growers of this crop. He first tried the treatments advocated by expert consultants without any very marked success. Then he adopted his humus methods and this trouble disappeared; moreover, he has now grown a crop of tomatoes in his houses for eight consecutive years without changing or sterilizing the soil and, so far, can detect no falling off in yield or quality. This is usually considered by experts to be impossible.[2]

The high quality of the Iceni produce is famous wherever it is known, and Captain Wilson has demonstrated very clearly that quality, even today, has a very definite commercial value. This is often contested by the users of chemical fertilizers who argue that successful competition under modern marketing conditions involves putting quantity before quality. Recently during one of the periodical bad slumps in the demand for cabbages, brought about by a seasonal glut of that vegetable, Captain Wilson sent a van-load of cabbages to the market of a town in the Midlands. When the name on his van was spotted, it was quickly surrounded by buyers, and the whole load was easily disposed of. Afterwards, rather than return empty, the van loaded up with the unsaleable cabbages sent, on the same day to the

[1] Made on the biodynamic principle.

[2] This result has been confirmed by several other commercial tomato growers. Again and again they have reported the complete disease control by the use of compost only. Incidentally, the tomato is one of the plants that thrives best on compost made from its own waste materials.

same market, by other growers. These were taken back to Surfleet and added to the compost heaps.

Plant pathologists when visiting Surfleet have jokingly called it 'a most uninteresting place', because it is, as they put it, 'so boringly healthy'. If the primary cause of disease in plants (as in man) is faulty nutrition, then Captain Wilson can claim that his organic treatment supplies his crops with a balanced diet.[1]

We must now turn to the question of parasites proper. These can be divided into those that live in the soil and attack the roots or stems of plants, and those (such as insects) that live outside the soil, and usually attack the leaf structure.

We have seen that the beneficial effect of compost in controlling virus or fungus disease is probably due to increased fungal activity, and the part this plays in providing the plant with a balanced diet. I now hope to show that there exists an equally plausible reason to explain the decrease in attack by parasitic grubs, which so often follows humus treatment. One of the most interesting examples of the kind is the effect on eelworm which has recently been noticed in South Africa and Ceylon following applications of compost. An account of this appeared in the *Rhodesia Herald* of 4 September 1942 as follows:

'Some years ago Mr. S. D. Timson, Assistant Agriculturist, noticed a garden in which the vegetables were strong and healthy and the flowers bright and vigorous. He was surprised to learn that three years earlier cultivation had been almost abandoned because of the heavy infestation of eelworm. The excellent conditions he saw followed a good dressing of compost.

'He immediately began to observe the results of compost in regard to eelworm, make practical tests and induce farmers to experiment. Once the inquiry was begun evidence began to pour in.

'At Darwendale Mr. O. C. Rawson had applied 5 tons of compost per acre to infested tobacco land. In the first year there was a reduction of eelworm, and in the second year, without a further application, the eelworm disappeared. Other tobacco farmers began to report similar experiences. . . . The compost, of course, was applied for its fertilizing value and the consequences on the eelworm population were a surprise.

'It now seems that the same effect has been noticed in other countries where compost was being tested out. In Ceylon members of the Department of Agriculture were able to report last year that the most promising method of ridding the soil of eelworms was one they

[[1] On Captain Wilson's retirement, this estate was sold.]

had only recently discovered. It had been found that if large quantities of organic material, such as compost, green manure, or cattle manure, were added to the soil the population of eelworms was greatly reduced. Examining into the causes of this effect, their conclusion was that it was due to the increase in the soil, following the addition of organic matter, of organisms like fungi, and insects, and other nematodes, which preyed on the eelworms. . . .'

This result would not have been a surprise if the machinery for co-ordinating the results of research in different branches of science were less rusty. No less than fifty-six different species or forms of fungi have been recorded by mycologists as 'subsisting by the capture of motile animals'.[1] The first account of this phenomenon appeared as early as 1888 (Zopf). Of these fifty-six forms, twenty-five are known to capture and consume nematodes, twenty-three kill and eat amoebae, five live upon rhizopods and the three remaining species are aquatic.

The methods by which these fungi trap their prey vary. The majority of the nematode-eating species form loops or bales of mycelium, which also usually excrete an adhesive substance. When an eelworm crawls through these bales, they close round it, holding it captive despite its violent struggles which may last for as long as two and a half hours. When the captive becomes inert, the fungus bores through the skin of its prey, making a narrow penetration. A globous body is then formed by the penetrating mycelium, which increases in size until in an hour or two it has filled a transverse section through the animal's body. Lateral elongated branches then grow from this, passing between the captive's internal organs, until they fill its entire length. This causes paralysis and death of the eelworm, after which the fungus consumes all the internal fatty material of its body, leaving the cuticle untouched.

Nematodes appear to form the normal diet of some forms of predacious fungi, other forms, it seems, eat such active prey only in the absence of other food. Comparatively recent investigations have disclosed the fact that many more forms of fungi are predacious than has been supposed. These forms have all the appearance of non-predacious species until brought into close proximity with their mobile prey. Then, and only then, do they develop the catching apparatus.

Many forms of these fungi are normally found in rotting vegetable matter and the dung of animals. An examination has also revealed their presence in very large numbers in compost.

[1] Charles Drechsler, 'Predacious Fungi', *Biological Review*, XVI, 1941, pp. 278 and 290.

Besides capturing and eating the active larvae of nematodes parasitic to plants—such as wheat cockle and sugar-beet eelworm, and the root-knot parasite of pineapple—some forms also capture and consume the larvae of many animal parasites. These larvae, which are present in the dung of affected animals, are quickly snared and killed by the fungal species concerned.[1]

This probably explains the greater liability to attack by parasitic worms to which cattle and sheep are subject on worn out pastures deficient in humus, for it is unlikely that predacious fungi can thrive in the absence of humus, any more than other soil forms.

Whether or not these groups of fungi are among those concerned with the production of growth-promoting substances is not known. It would be rash, in the absence of precise knowledge, to assume that they are not. If they *are*, then the deliberate destruction of eelworms by chemical means, may, even where successful, do more harm than good by destroying the natural food of these beneficial fungi. The normal presence in dung of both nematodes and the fungi which eat them, certainly suggests that a balance between them exists in nature, particularly as some nematodes and rhizopods appear also to eat fungi, if not first eaten *by* them. These organisms provide an instance of the importance of the study of soil ecology, and of the possible evil consequences of upsetting nature's balance of species.

' "There is no human being who is not directly or indirectly influenced by animal populations, although intricate chains of connection often obscure the fact . . . not only do animals have this influence on man, but man has an increasing power over the fate of the animal populations that still throng the world." It seems that considerable caution should be exercised before "ordering the destruction of a species on the chance that it may be doing harm to human interests."[2]

'We cannot condemn the prairie gopher, the Australian rabbit, the African termite, or the locust, without first discovering to what extent, if any, some action of man has upset an equilibrium in animal ecology, and therefore indirectly in plant ecology, and so affected the conservation of vegetation, soil and water.'[3]

Man is certainly responsible for the devastations caused by the Australian rabbit, though in this case he upset nature's balance by the introduction, not the extermination, of a species. As for the African

[1] See Tech. Ref. A, p. 159.

[2] Bureau of Animal Population at the University of Oxford—Annual Report (1936–7) quoted from *The Rape of the Earth.*

[3] Jacks and Whyte, *The Rape of the Earth*, Faber & Faber, London, 1939.

termite, Captain Moubray has recently shown that it can serve a very useful purpose. He uses large quantities of compost on his citrus farm in Rhodesia, and he has described (April 1942) how he has recently been able to make use of the termite in providing an additional source of food for his trees.

'We have a small box-making plant,' he writes,[1] 'and until recently sawdust was looked upon as a waste product. Now it is spread on the grove, where the termite or white ant of the insidious variety has made its home. This insect has the property of fixing atmospheric nitrogen, which enables it to digest cellulose and convert it into protein, which is a valuable source of nitrogen.

'We used to look upon the insidious termite as a pest and sawdust as waste. By combining the two we make a valuable product. The termites do no harm to the trees. The American representative of Cyanogas pays me regular visits, and has been watching this particular experiment with great interest.'

In a later letter Captain Mowbray writes: '*Re* termites, I now think the correct explanation is as follows; the termites take the sawdust underground and use it in the manufacture of their mushroom beds; this particular termite makes nests about the size of a cricket ball, only a few inches below the surface. When finished with, these beds decompose, nitrification takes place, and the nitrogen is available to the roots of the orange trees. As regards eelworm in tobacco, our experts appear to be fairly satisfied that this pest is now controlled by good applications of compost. Up till recently it was generally accepted that once lands became badly infested they were lost to tobacco forever. Many more have repeated Mr. Rawson's and Captain Timson's trials with the same results. I found this out on tobacco years ago, but people then treated it as a joke.' (Note: the expert advisers on sugar-beet and potato eelworm in this country still treat it as a joke.)

Whenever pests appear in unmanageable numbers, it is probably safe to assume that nature's balance of species has in some way been upset and in most cases it would be profitable to suspect that the root of the trouble is man's mismanagement. For example, there is some evidence for attributing the increase in insect pests, which has been reported in some areas, to the wholesale destruction of hedges, undertaken in the interests of mechanized farming, for with the hedge has gone the shelter for the small birds who prey on the insects. Similarly, any soil treatments which tend to inhibit fungal activity

[1] *Compost News Letter No. 4.*

are probably a direct or indirect cause of the increase in many different forms of parasitic larvae. In Hawaii successful control of root-knot parasite in pineapple has been achieved by treating the soil with large quantities of organic matter applied with the deliberate intention of encouraging predaceous fungi.[1] Farmers on the eelworm-infested potato lands in Lincolnshire might do well to follow this example.

In addition to the undoubted presence of predaceous fungi in such substances as compost and dung, other factors concerned with the balance of the soil population are probably called into play as a result of organic soil treatment. There are indications, for example, that attacks by wire worm on newly ploughed-up pasture are noticeably less severe if the grass receives a heavy dressing of farmyard manure before ploughing.

Not the least important effect of replacing chemical fertilizers by farmyard manure or compost, is the increase which this brings about in the earthworm population. Darwin found that the weight of worm casts deposited on the surface of ordinary field soil in good heart may exceed ten tons per acre per year. The importance of this and its direct bearing on plant nutrition, becomes clear when you consider the estimates of other experts, notably Dr. L. C. Curtis of the Connecticut Experimental Station, that worm casts contain five times more nitrogen, seven times more available phosphate, eleven times more potash, and 40 per cent more humus than is normally to be found in the top 6 inches of soil, and this is not their only contribution to soil fertility.[2] The late Sir Bernard Greenwell attached great importance to the part played by earthworms in pest control and in the production of healthy crops generally. In a paper which he read to the Farmers' Club in 1939 he stated:

'I am afraid very few of us realize what a good friend this little fellow is to the farmer, and if we can only increase the population of the earthworm in the soil he will do a lot of our deep cultivation for us and aerate the soil gratis. Where we manured our grassland with artificials, we found the worms disappeared, but the following year a compost was applied made from town rubbish mixed with dung, and immediately the worm casts reappeared. It is a known fact that the nomad tribes in Central Africa always pitched their camp on ground covered with worm casts as they found that this was the best grazing. . . . I am certain that the fertility of the soil is bound up with

[1] See Tech. Ref. B, p. 160.

[2] A list of some of the further beneficial activities of earthworms will be found in Tech. Ref., p. 112.

organics which are a great encouragement to the worm. There is very little doubt that he is a scavenger and if he disappears you will find his place taken by the leatherjackets and other insects detrimental to the crops.'[1]

When it is remembered that as many as eight million earthworms may be found in a single acre, and that their burrows are capable of penetrating the subsoil to a depth of five or six feet, it is clear that their presence must have an appreciable effect on soil aeration, and thus indirectly on humus formation.[2] Faulty aeration, which interferes with humus formation underneath the turf, is perhaps the commonest cause of infertile grassland.

Sir Bernard Greenwell found that he could obtain an improvement in poor pasture by subsoiling, equivalent to that normally obtained by dressings of basic slag. It will be remembered that some of the Wareham experiments suggested that the improvement in tree growth brought about by slag was due less to its phosphate content than to its indirect action on soil fungi, including no doubt those responsible for humus formation. It seems fairly clear that an increase, however induced, in that part of the soil population beneficial to plant life may well be accompanied by a decrease in others of a parasitic nature.

We come now to consideration of the claim that compost treatment also confers a resistance to attack by *insect* parasites. Here the reason is more difficult to understand. At first sight one would have thought that the healthier a plant, the more appetizing would it be to such pests. The recorded cases of such resistance following applications of compost, are, however, too numerous to ignore. My own experience extending over several seasons has convinced me that crops of the cabbage family, if grown from seedlings raised on humus rich soil, and transplanted on to land treated with compost, show markedly greater resistance to caterpillar and aphis attack than those grown with artificials or even with ordinary farmyard manure. I have also grown onions in soil rich in humus which produced an abundantly healthy crop in a season when practically the whole onion crop of the surrounding district was destroyed by onion fly.

Howard records the following interesting experience with fruit

[1] It has also been suggested (Dr. Joad, in a Brains Trust session) that the increase in such pests as wireworm, and leatherjacket, may be attributable to the speed of the multifurrow tractor plough, which does not allow the grub-eating birds sufficient time to secure their prey before the fresh-turned soil is covered up again by the next furrow. This is very plausible, and there is probably something in it.

[2] See Tech. Ref., p. 112.

trees found growing in the garden of his house at Blackheath when he bought it in 1934. He records that this garden 'was completely worn out through no fault of the previous owner. It was a veritable pathological museum—the fruit trees, in particular, were smothered with every kind of blight. Steps were taken to convert all the vegetable wastes into humus with the help of stable litter. Even after one year the pests began to retreat. In three years all had disappeared, the woolly aphis on one apple tree being the last to leave. During this period no insecticides or fungicides were used and no diseased material was ever destroyed. It was all converted into humus.'

Every year that passes brings an ever-increasing number of carefully recorded results of a like nature, for organic farming and gardening is daily becoming more common. Among the outstanding features of all these reports are the disappearance of disease, the increased size, and improved germination of seed, and the increased resistance to drought and pests, all following upon a change from orthodox cultivation to compost only. In none of these cases has a diminution in yield resulted—usually the reverse.[1]

I have evolved my own theory to explain the effect of compost on insect pests, which I advance here merely as an hypothesis. In watching caterpillar and aphis attack I have noticed that both compost- and chemically-grown plants will be attacked, but whereas the plants grown with chemicals (if left unprotected) will, in a bad season, be stripped to the rib, those grown with compost (with the exception of an occasional isolated plant) are only slightly damaged and quickly recover. This suggests two things: first, that the vigour of growth is so strong in the case of the compost-grown plants, that they are able quickly to repair the ravages of the pest, and secondly, that the compost-grown plant may be more nourishing and the parasite consequently satisfied with less. The evidence on which I base this theory will be found in the next section.

[Other explanations have been offered. The greater strength of the cell wall in compost-grown plants—which may also explain their greater frost resistance—may offer more resistance to insect mandibles. The different smell (or possibly radiations) of compost-treated soil is another hypothesis. It has also been suggested that pests are nature's plan for removing the unfit.]

[1] These results are now being collected and checked by the Soil Association. [See p. 370.]

II

ANIMALS

We now come to the second claim made on behalf of humus, namely that the health of the crops grown with it can be transmitted to animals.

This is not easy to prove, but as an example of the instances given, I will briefly summarize the experience of Mr. Friend Sykes,[1] farmer of a 700-acre all-organic farm. For many years Mr. Sykes farmed some of the richest land in the country, where he built up famous prize-winning herds of Friesian cattle and Berkshire pigs, yet when the accredited milk scheme was introduced, and he was asked to lead the way by having his herd tested for tuberculosis, 66 per cent of his cows reacted. As can be imagined this caused him many heart searchings. For the conclusions he reached, and the reasoning by which he arrived at them, readers must study his book,[1] but the practical upshot of them was that he decided radically to alter his system of farming, and abandon orthodox methods. He sold his rich land farm (at £100 per acre) and in 1936 bought 750 acres of downland in Wiltshire over 800 feet above sea-level and worth only £4 an acre.

This land was so poor that it was said to be incapable of growing grain crops, and the herbage was so scant that the whole area, when he first moved in, was incapable of sustaining fifty head of cattle, and heavy purchases of feeding stuffs had to be made to keep them alive. Disease of every sort soon showed itself: contagious abortion, Johne's disease, mastitis, tuberculosis.

He decided to plough up the whole 750 acres, to rely thereafter solely on home-grown food, and to attempt to re-establish fertility without recourse to any purchased artificial fertilizers. He records the result in the following words: 'After seven years of heart-breaking toil with the added difficulties of wartime conditions thrown in for luck, we have (touching wood) (a) almost completely rid the farm of disease, (b) built up a large herd of attested dairy cattle tubercle-free for over four years now, and of a soundness of constitution to all critical appearances such that no expert would believe that any scourge had ever visited the farm, and (c) as each succeeding generation of young stock is born we have unmistakable evidence of still greater stamina and endurance.'

[1] Friend Sykes, *Humus and the Farmer*, Faber & Faber, London, 1946.

This remarkable achievement was brought about by a combination of ley farming (four years seeds, four years arable), heavy cultivation including sub-soiling, and intensive controlled grazing by cattle, sheep and horses. Beyond this heavy sheet composting the land received nothing, yet the farm, once so poor, is now self-supporting, carries 250 head of cattle, and grows heavy crops of roots and grain, which are as healthy as the livestock. In 1944 the wheat crop reached seventy-two bushels per acre, and this on land said to be unsuitable for wheat both on account of quality and altitude.

The farm is still on the up grade, and now that Mr. Sykes is composting all his straw by the method of open yards and a mechanical muck shifter he expects to be able to carry a much larger head of livestock still.

Now this conversion of derelict and infertile soil into something much above the average for the best, through no agency other than the dung and urine of the livestock *fed on the farm's own produce*, while grain and milk are sold off the farm, cuts clean across the orthodox view of what is possible. By all the rules a serious mineral deficiency should have taken place. The reverse appears to have happened.[1]

If animals fed on compost-grown food *are* healthier than those fed on the same foods grown by other means the explanation is probably to be found in a combination of many factors, and to single out any one of them is perhaps to fall into the error of fragmentation. Nevertheless, since vitamins have already been singled out for special investigation in recent research on this subject, and since they undoubtedly do appear to be one of the factors concerned, some of the available evidence in connection with them must now be given.

In Madras McCarrison found that grain produced with farmyard manure contained more vitamins than that grown with minerals.[2]

A striking confirmation of this finding in the case of a particular

[[1] See Part II, p. 207.]

[2] *Journal of Medical Research*, XIV, 1926, p. 351. The following results of one of his rat-feeding experiments demonstrates this:

Feeding	*Percentage gain in bodily weight*
Basic ration plus stable-manured wheat	114
Basic ration plus chemically grown wheat plus vitamin supplement	104
Basic ration plus chemically grown wheat alone	89

The two wheat crops were grown on adjoining plots. (*Memoirs of the Department of Agriculture in India*, IX, 4, 8927. Quoted, Pfeiffer.)

vitamin—B—is to be found in an experiment carried out by Rowlands and Wilkinson of the Knightsbridge Laboratories.

The following account of their experiment appeared in *Compost News Letter No. 4* and was abstracted from the *Biochemical Journal*, XXIV, 1, 1930.

'Mr. Rowlands, having noticed that pigs grown on home-grown barley and wheat did better than pigs on bought barley and wheat, decided to try the effect of artificial manure versus natural manure. The crop tested was, not barley or wheat, but the seeds of clover and grass; and the experimental animals were rats. Two groups of rats were fed respectively on the seeds from the artificially manured field and the naturally manured field; and the upshot was that, on the seeds of the naturally manured grass, the rats grew nearly twice as well as those on the artificially manured grass. Both preventive and curative tests were made.

'Here are the details:

Preparation of field	1925	Cabbages grown.
	1926	Potatoes grown. Wheat sown in autumn.
	1927	Grass seeds with a small quantity of the different clovers sown amongst the growing wheat. After harvest, the aftermath was lightly grazed.
	1928	Field cut for hay and grass. In autumn the field was divided and half was manured with natural manure and half with artificial thus:

Natural		*Artificial*
Pig manure mixed with straw 20 loads per acre. How the pigs which produced the manure were fed: 50 per cent middlings (i.e. ground bran with adherent endosperm), 40 per cent barley meal, 10 per cent of the following mixture: meat meal, rye and wheat embryo (*rich in vitamin B*), bone meal and cod liver oil.		20 cwt. basic slag and 3 cwt. Kainite (sulphate of potash, salt and magnesia) per acre.
	1929	In the spring 1 cwt. sulphate of ammonia per acre.

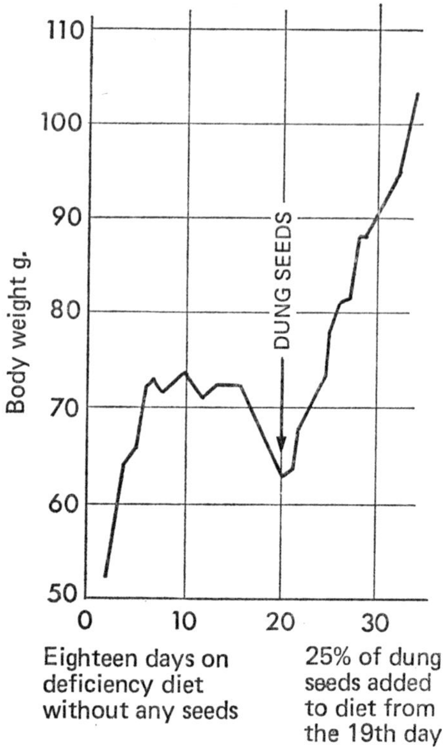

FIGURE I

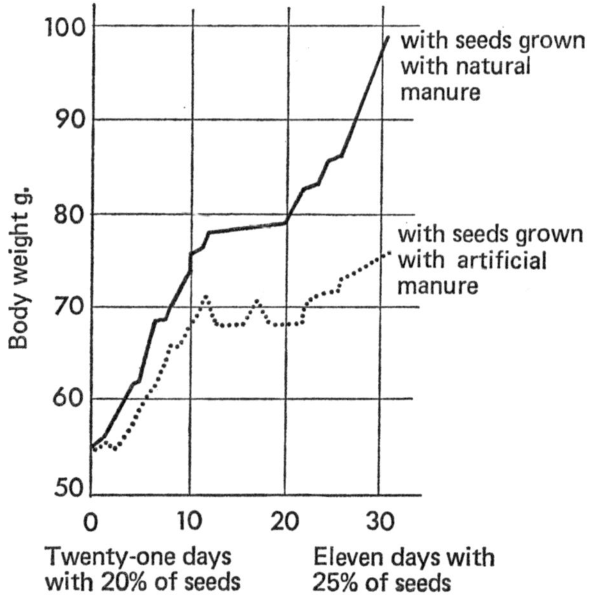

FIGURE 2

Harvest

'In July 1929 the crops were cut on both parts of the field; harvested in separate barns, and threshed.

Analysis of Grass and Clover Seeds (percentages)

	After natural manure	*After artificial manure*
Protein	12·70	11·70
Moisture	5·80	6·00
Fibre	19·40	21·60
Ash	9·00	10·50
Phosphorus	0·44	0·33

Preventive Tests

'The rats were given Professor Drummond's "B deficiency diet" which consists of rice starch, caseinigen, salt mixture and cod liver oil, and with this, one lot got 20 per cent of "Dung Seeds" (seeds grown with natural manure) for twenty-one days and thereafter 25 per cent of "Dung Seeds" for eleven days, whilst the other lot got the same weights of seeds grown with "artificials", for the same periods. The result is shown in Fig. 1 (p. 130).

Curative Tests

'Both lots of rats to be tested were given the "B deficiency diet", with *no* seeds, for eighteen days, at the end of which all were losing weight. Then to the diet of one lot 25 per cent of "Dung Seeds" were added, and they at once began to grow normally. (See Fig. 2 (p. 130).)

'To the diet of the other lot 25 per cent of seeds grown with "artificials" were added: they did not improve, but got steadily worse. Three days later they looked as if they could only live a few hours. "Their condition was one of typical vitamin B deficiency; they were wasted, hunched and shedding hair." Then Professor Drummond saw them and suggested changing from seeds grown with "artificials" to "dung seeds". The immediate result was that the rats rapidly recovered and showed normal growth. (See Fig. 3 (p. 132).)

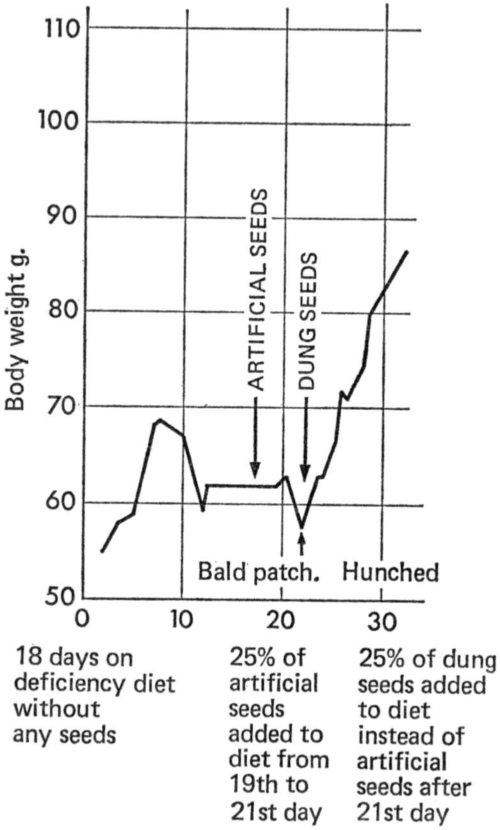

FIGURE 3

'The authors observe "It would seem that a plant may absorb vitamin B from the land, and that the vitamin B content of any food may be dependent upon the amount of this vitamin in the land." The earlier view that "the vitamin B is manufactured entirely in the plant and stored in the embryo" thus seems to be incorrect. Then they make this shrewd remark: "One worker states that in his experimental work a certain food contains vitamin B; another worker, using the same type of food, differs entirely in his findings; and we consider that the food that is being tested is not the only point which is of importance. If these results are correct they have a considerable bearing upon agriculture and nutrition."

'A further detail: They found that the dung of the pigs which they employed contained vitamin B.[1] (They extracted it with slightly acidified 50 per cent alcohol, concentrated the extract by distillation in vacuo, and then fed the concentrate to rats losing weight on the "B deficiency diet". The rats began to grow at once.)'

This experiment dealt with vitamin B only. Very probably it could be extended to include others. A similar result, but without an attempt to isolate the cause, is recorded of the late Dr. Rowlands:

'In his case he took an acre of land that had never had a crop on it, divided it into two half-acre plots, ploughed, harrowed it, and made a seed-bed of the whole acre. One half-acre was heavily fertilized with cow dung, the other half-acre with chemical fertilizer—both plots seeded down with the same cereal—barley, I think it was—both plots had good crops with strong straw and well-filled heads, the seeds were kept separate and again two groups of rats from the same families were divided in equal halves—the one group fed from the chemically fertilized plot showed every evidence of malnutrition—those fed from the farmyard manured plot had a perfectly normal healthy growth.'[2]

As against such experiments as these I must in fairness record certain contrary evidence. First there is the increasing development of soilless culture, or hydroponics as I believe it is called in the USA. It is usually described as growing crops in a nutrient solution, either in water tanks or in sterilized sand or gravel, though it is sometimes forgotten that water or sand when exposed to the air does not remain sterile. Considerable progress has been made in this method. It has even reached a commercial scale in some places. Satisfactory crops have resulted, and Professor R. H. Stoughton, who is engaged in carrying out investigations in the process at Reading University, has recently recorded that in order to determine the nutritional value of crops so raised, 'chemical analysis of the carbohydrate, protein, inorganic constituents, and vitamin C content were carried out. No significant differences could be established between plants grown in gravel and those grown in soil.'[3]

The first question, of course, which springs to the mind is 'what kind of soil?' If chemically treated soil, then one would not expect to find much difference, but apart from this my reactions to these investigations are, firstly, that the results will be inconclusive until

[[1] A good example of transference within the food chain. See also Appendix, p. 366.]

[2] Personal letter to the author from Mr. J. Tustin, late of United Dairies.

[3] *Nature*, CL, 3797, 8 August 1942.

feeding tests have been carried out such as those undertaken by Dr. Rowland. It is very possible that the difference in nutritional value between humus and chemically grown plants is not, as yet, detectable by chemical analysis. It must not be forgotten that until very recently vitamins defied analysis.[1] Secondly there is no indication at present that plants raised by soilless culture can continue to reproduce for consecutive generations. It is customary for the seed used to be raised in soil.[2] Thirdly, plants grown in such very controlled and artificial conditions present a very different picture from those growing in open competition in the field. The danger of basing deductions exclusively on laboratory experiments when dealing with such natural phenomena as mycorrhizal association, for example, has already been pointed out. Of more importance to our subject, therefore, is the evidence of Sir John Russell, FRS, late director of the Rothamsted Experimental Station. In 1939 he wrote as follows:

'We have searched diligently for evidence that organic manure gives crops of better quality than inorganic fertilizers, and so far our experiments made jointly with the Dunn Nutritional Laboratories at Cambridge, have all given negative results. No difference has yet been found.'

How can such results be reconciled with those of Dr. Rowland? I am not in possession of sufficient data to hazard an answer, but several lines of inquiry occur to me. Firstly, in this case also, no evidence is forthcoming that long-term feeding tests were undertaken. Secondly, I would like to know the manurial treatment which the soil received in the years immediately preceding the experiments, for the effects of both organic and chemical manuring are cumulative. If the operative factor is a question of fungal activity, it is obvious that this must be so. Thirdly, I should like to know the form of the organic treatment given, for, as we have seen, not all treatments called organic do stimulate the soil population in the desired manner.[3]

[1] 'About twenty years ago I was induced to preside, during the annual meeting of the British Association, over a joint conference of its Agricultural and Medical sections arranged to discuss vitamins. The subject was handled with considerable scepticism by the medical participants in the discussion. I well remember, prior to its taking place, I wrote to several well-known physicians asking them their opinion on the value of vitamins. Their replies were almost without exception hostile to their alleged efficacy, and in some cases derisory. And yet who today would dare to challenge the therapeutic value of these "accessory food factors"?' (Viscount Bledisloe, 1940.)

[2] Mrs. Hilyer records in her book *Hydroponics* having successfully raised second-generation peas and potatoes but her methods are not entirely inorganic.

[3] A recent visit to Rothamsted revealed the fact that no tests have been made with properly made organic compost. (1948.)

It must be remembered that the inorganic enthusiasts tend to overlook in their claims that application of inorganic chemicals to a field crop does not so much test the effect of chemical nutrients as such, as of chemicals plus humus. This is why the humus enthusiasts demand long-term tests. Their view is that it is the presence of humus which softens the effects of inorganic chemicals, and the presence of inorganic chemicals that prevents compost treatment being fully effective. In other words, while it is only when the humus content of the soil becomes seriously lowered that the harmful effect of inorganics becomes *fully apparent*, feeding value nevertheless is probably affected long before this stage is reached. The only short-term tests, therefore, which could possibly provide a true comparison would be one undertaken with subsoil; in the one case with inorganic nutrients added, and in the other case with the addition of humus. I believe it is a fact that the seed used on the famous Broadbalk field at Rothamsted is invariably fresh seed, imported each year. This must weaken any nutritional argument based on the reaction of this seed to soil treatment, for since the effect of both organic and inorganic treatment is cumulative, it becomes more and more evident with each succeeding generation. One of the reasons for this is no doubt that continued applications of inorganic chemicals appear to have a serious lowering effect on the fertility principle, while organic treatment, as already noted in the case of sunn hemp, stimulates it. A similar result has been noted with clover by Mr. R. G. Hawkins of Braintree who writes (May 1942):

'For some years now I have inspected crops of Essex red clover and I have noted that the yield of seed is invariably higher on those farms which keep stock, so that the land receives a periodic dressing of dung. The difference is most pronounced in those years when clover seed is generally a poor crop. . . .'[1]

The factor responsible for lowering reproductive powers, is probably the same as that affecting feeding value. In both cases it is reasonable to assume that it is almost certainly a vital factor. The argument that living forces are unnecessary to living organisms, and that these require nothing for health, growth and vigour that could not be supplied by synthetic chemical products, will remain unconvincing until such time as scientists succeed in creating life itself. Dr. Innes Pearse maintains that the relationship of humus to the seedling, is exactly comparable to that of the placenta to the fœtus. In any event, the mere fact that such startling discrepancy of view as that cited

[1] Quoted from *Compost News Letter No. 3*.

above can co-exist among scientists is in itself a sound indication of the importance of carrying out further tests.

There are very many indications now available contradicting the Rothamsted results and they are not confined to rats.

In the course of feeding trials carried out by the late Sir Bernard Greenwell at Marden Park in Surrey, in 1939, 'the effect of a grain ration raised from fertile soil was compared with a similar one (purchased on the open market) on poultry, pigs, horses and dairy cows. In all cases the results were similar. The animals not only throve better on the grain from fertile soil, but they needed less—a saving of about 15 per cent was obtained. The grain from fertile soil was found to contain a satisfying power not produced by ordinary produce. But this was not all; resistance to disease markedly increased. In poultry, for example, infantile mortality fell from over 40 per cent to less than 4 per cent. In pigs, troubles like scour disappeared. Mares and cows showed none of the troubles which often occur at birth.'[1]

The possibility that grain grown on fertile soil might result in a quantitative saving is an important point, particularly in war-time. If a wide adoption of humus farming were to result in a 10–15 per cent reduction in animal feeding stuffs throughout the country, a large amount of shipping space would be saved, or alternatively more livestock could be kept. These are worthwhile considerations in themselves, quite part from the economic benefit that would accrue from any general improvement in the health of our livestock, were this to result. I might remind you here how Howard found that his oxen in India, fed on compost-grown food, failed to contract foot-and-mouth disease even when 'rubbing noses' with infected animals.[2]

In connection with this Lord Portsmouth told me an interesting story. A few years ago he approached the heads of the Government Veterinary Department, whose responsibility it is to decide on the policy to be followed during epidemics of such infectious diseases as foot-and-mouth disease. At present, as is well known, the policy in the case of this disease, is to slaughter not only all infected animals, but also all contacts, at the same time prohibiting any movement of stock in an area fifteen miles around the site of the outbreak. Anyone who has studied the outbreaks in recent years, cannot fail to have been struck by the way in which farms next door to the one where an outbreak has occurred, so often escape infection, while others, more

[1] Howard, quoting Sir B. Greenwell's paper read to Journal of Farmers' Club, 1939.

[[2] See note at the head of this chapter.]

widely separated, succumb. It is difficult to reconcile this with the official theory that the spread of infection is due to birds, or ground vermin.

Lord Portsmouth, with his personal experience of the effects of humus farming, suggested an experiment to these Veterinary Authorities, to be tried when foot-and-mouth next occurred in a suitable locality, such as the Isle of Wight, where a whole area could be effectively isolated. His idea was that instead of adopting the slaughter policy, they should prohibit any movement of stock to or from the island, and let the disease take its course, carefully noting the farms that escaped and comparing the system of farming in practice there, with that on the farms to which the disease spread. The 'vets' 'threw up their hands in horror' and said: 'Do you suggest that half Europe is badly farmed?'

'Yes,' said Lord Portsmouth, 'just that.'

They then drew his attention to a large wall map. It covered the Continent from the Baltic to the Mediterranean, and from the Atlantic to Russia. It was studded with small flags, each flag representing a confirmed outbreak of foot-and-mouth disease.

Lord Portsmouth told me that he looked from this map to the 'vets' in astonishment. 'Doesn't that map show you something?' he asked.

'Only,' was the reply, 'that the disease is very prevalent.'

Lord Portsmouth then told me that he had an intimate knowledge of some parts of southern Europe, and had travelled large tracts of it on foot, including the Balkans. He said that from his personal knowledge of the farming methods undertaken in this mountain area, he was able to notice that very often the flags stuck into the map stopped where traditional peasant farming began. He pointed this out to these authorities, saying, 'Do you mean to tell me that birds flying south stop when they get to this valley or that region?'[1]

I asked him what reply was made to this, and he said: 'Oh! They just thought I was mad!'

There is nothing harder than to introduce a new idea into the official mind.

A further contribution to the indications concerning animals and artificials, is contained in the following letter to the *Spectator* of 17 October 1941, by Dr. Sanderson-Wells, Chairman of the Food Education Society.[2]

[[1] See note at the head of this chapter.]
[2] T. H. Sanderson-Wells, MBE, MD, FRCS.

'A dig-for-victory plot on the edge of a golf-links was limed, planted and treated with artificials. Luxuriant heavy-green cabbages, sprouts and other vegetables resulted. To increase the family meat ration, part of this crop was fed to rabbits, who ate without relish, became apathetic and smelt unpleasant. When later grass mowings were substituted the rabbits ate voraciously and became vigorous and sweet-smelling.

'A correspondent writes: "Cabbages and sprouts grown too fast with nitrate and phosphate are a curious 'wrong' colour. If over 50 per cent of the greenstuff given to rabbits is of this sort the rabbits die. Permanent pasture dressed with phosphate produces a luxuriant field. If the phosphate goes beyond a certain point the field takes on an unnatural green, and is deserted by wild rabbits."

'Salesmen use this fact as a recommendation. One told me: "Use any soluble phosphate fertilizer and keep the rabbits away." Another said: "Use enough nitro-chalk and you will get big greens that rabbits will scarcely touch: if they do, they die." Animal instincts may be sound guides to food values, which are actually soil values, because food is nothing more than the "conveyancing agent" or "agent of transfer" of the soil's qualities into the bodies of man and beast; land in good heart supporting bodily health, vigour and stamina; poor unbalanced soils producing ill health and debility. For instance, the liking of birds for hips, haws and many other hedge-row fruits has led to the discovery that these contain high concentrations of mineral salts, vitamins, and other essential food elements.'[1]

The soundness of this idea that food is an agent of transfer for the qualities of soil has been demonstrated to me many times by my own experience in the raising of young pigs. Pigs bred under modern housing conditions (defined by someone as 'cold and concrete'), are very prone to the disease of white-scour when they reach the age of about one month. If the attack is serious it can cause considerable financial loss, even if it does not actually kill the pigs. The textbooks give the cause as lack of iron and recommend dosing with some iron preparation such as Parrish's Food; feeding such weeds as chickweed (which is rich in iron), or, as a third alternative, taking up pieces of turf and giving these to the young pigs.

I have made many experiments in connection with the curing and prevention of this trouble. From the turf remedy I tried experiments with ordinary soil from arable fields. It was not long before I found

[1] Reprinted in *Compost News Letter No. 3*.

that soil gathered from a field rich in humus, where no chemicals have been applied, was quite as effective as turf, curing the pigs within forty-eight hours. Whereas soil from exhausted land, or land treated with chemicals had no effect in curing the disease. I also noticed that young pigs running in the open on good pasture, provided it was not too hard for them to rootle (as for instance in hard frost, or very prolonged drought), never suffered from this disorder. It is never a menace to my herd now under any conditions, even in long spells of severe winter weather when the ground is covered with snow and the pigs have to be entirely housed up. Under such conditions I no longer wait for the first sign of scour, but regularly collect the soil of *fresh* mole hills, newly thrown up above the snow, on land I know to be fertile. Collected daily, this soil is friable in the hardest frost, and it is equally good in very wet weather, for it is never sticky. The pigs eat it voraciously in incredible quantities starting when about a week old. I sometimes add a little chalk to it, which the pigs seem to like.

This experience ties up in an interesting way with the research work of Baker and Martin[1] and brings us back once more to the role of the fungus. Baker and Martin have shown that a group of fungi indigenous to the caecum of the rabbit, guinea-pig, and horse is responsible for the breakdown of cellulose ingested by these animals, and their research work on the behaviour of these fungi *in vivo* has provided strong evidence for the view that their relationship to the host animal is a symbiotic one. (Some of these organisms incidentally are also indigenous to the human mouth.)

These investigators have not experimented with the pig so far as I am aware, but it is reasonable to suppose that the breakdown of cellulose in the pig's digestive tract is very probably also dependent upon the presence of fungi. It is also reasonable to suppose that the pig's gut, at birth, is more or less sterile. This being so, before he can cope with green food or cereals, he must presumably inoculate himself with appropriate fungi. If this is, in fact, what occurs, it explains the pig's very early appetite for soil. It also explains why soil heavily dressed with artificials is not effective in stopping the indigestion which causes white-scour at about the age when the pig normally begins to eat green food, since in such soil fungal growth is largely inhibited.

An alternative explanation, put forward by certain medical friends, is that the scour is bacterial in origin, and that some bactericidal

[1] F. A. Baker and Rollo Martin, *Zbl Bakt., II*, 1937.

agency is introduced with the soil. This explanation is possible of course, and the recent discovery by an American scientist of a soil organism possessing curative powers very much akin to penicillium lends colour to the theory, but I still lean towards my own hypothesis, more particularly since research, more recent than that of Baker and Martin, has provided evidence that the digestive tract of all mammals is the normal habitat of fungus flora, symbiotic in habit, whose function is not confined to assisting the digestion of cellulose, but is also concerned with making certain vitamins available. If the initial source of fungus infection should prove to be the soil, or even natural raw herbage, one is prompted to wonder whether the inability to digest coarse natural foods, such as raw salads, and wholemeal bread, which is sometimes exhibited by modern city dwellers who have subsisted for years on sophisticated, processed and tinned foods, may not be associated in some way with faulty fungal activity brought about through the prolonged absence from their diet of those same natural foods. At any rate, in the case of the pigs, all the recent research is consistent with my interpretation, which also seems to be supported by the well-known fact that when cooked potatoes are fed to pigs they must be washed before boiling, whereas raw potatoes can only be safely fed if the soil is left on them. It would appear from this that live soil enables pigs to digest raw potatoes (in moderation), while cooked potatoes not only do not require this aid, but the soil, when sterilized, appears itself to be indigestible.

Older farm workers have told me that they never remember scour in young pigs when '*they* were young', and one of the more noticing among them told me that in his view this was because in those days when roots such as mangolds were cleaned for the cows, the part trimmed off, with the soil attached, was always thrown to the pigs, and that in 'those days, these crops were always grown with muck'.

There seems little doubt that we can make much more use than we do at present of the information which animals themselves are very ready to give us in estimating quality in food. For instance, cattle are often better judges than the chemist as to whether or not land requires liming. In the winter of 1941–2 I chalked part of a meadow which the soil experts, after analysis, told me did not require it. The following summer, and again during the summer of 1943, the verdict of the cattle was very different; that part which received chalk was grazed as close as a billiards table. The rest was fed off only when there was nothing left on the chalked portion.

The Rev. Willis Feast, of Booton, Norwich, records how a young farmer told him that he grew swedes, some with and some without artificials: 'He fed the "withouts" first, and when they were finished had the greatest difficulty in persuading his beasts to start eating the "withs".'[1]

The preference of cattle for mucked or composted pasture is very marked. Howard describes a case of this kind which he was able to observe personally, since the six fields in question were in front of his house at Haversham. 'All are first-class rye pastures,' he writes, 'with nothing to choose between them as regards soil, aspect or drainage. Nevertheless in 1941 the sheep and cattle, which had access to all six fields at the same time, consistently neglected one of them, the grass of which was allowed by the animals to grow at will. This particular field alone of the six had received a large dressing of artificials.'

One winter I noticed that the farm cats refused potatoes boiled for the pigs when these had been purchased from a grower who uses artificials, but that later in the season, when I started to use the small potatoes from our own crop, grown with humus, the cats ate them with avidity.

Stories such as these are indications, because they are reported from so many different quarters. They are not proof, because they have not been reproduced to order under strictly controlled conditions. I must confess, however, that my own personal taste entirely concurs with that of the animal. To me the flavour of vegetables grown with farmyard manure, and more particularly compost, is noticeably superior to that of the same vegetables grown with chemicals. It would be in keeping with nature's usually sensible arrangements if palatability were to prove an indication of nutritional value. Unfortunately, in so far as civilized man is concerned, such guides are no longer reliable, generations of faulty feeding having developed in us a liking for many things that do us little good, such as refined sugar and white flour. The natural choice of animals, however—even domestic animals—is almost certainly worth noting, and investigations based on it would probably yield interesting results.

The examples just given of animals obviously preferring compost-grown food, coupled with the experience of Sir Bernard Greenwell and others, that a smaller quantity of such food is required to produce a given increase in weight, certainly suggests that such food is more complete and balanced.

A doctor of my acquaintance holds the view that, 'It is a likely hypothesis that the appetite-satisfying qualities of compost-grown

[1] *Compost News Letter No. 3*. See also pp. 233–4.

food are due to a greater content of hormones and vitamins. Similarly when the vitamin content of such a food as bread, for instance, is deliberately strengthened by the addition of extra germ, provided that that is fresh and, except for the baking, raw, its power of satisfying hunger is notable.'

This doctor lives in Cheshire, a county where, before bread rationing, a special wholemeal bread was available. This bread was known as 'Fertility Bread'. It was made from locally grown wheat; about 66½ lb. of the latter with about 33½ lb. of raw wheat germ added, and was baked within thirty-six hours of the meal being ground.

He told me the following story as an illustration of the satisfying power of this bread.

'One rainy night, called to a distant and unknown patient, I returned from a fruitless inquiry to find a man standing by my car in conversation with my wife. She was saying—"I call a 4½d loaf expensive." And his reply was, "Well, I don't, Mrs. ——. You see I put it this way: I can make a good meal on half a fertility loaf and nothing more except the butter and perhaps a lettuce; and I am satisfied. That's 2¼d aside from the butter and so on. But if I must have white bread I want as much and more, and a piece of beef or a chicken's leg as well; and against I've done, my dinner's cost me more than double." '

This same doctor used a vivid analogy to illustrate the same theme. He likened an attempt to satisfy hunger on white bread, to eating Christmas pudding with intent to find the threepenny bit. 'It is not in the first helping, so you have another, after that you feel full, but though you do not really want to eat any more, you go on because you are still looking for something which is not there.' This analogy also expresses my theory of the susceptibility of chemically grown vegetables to caterpillar and insect attack. For example, a commercial tomato grower carrying out an experiment recently between a tomato house treated with compost only and one treated with compost plus chemicals noted a greater *number* of white-fly in the compost houses but *no* visible injury to the plant or staining on the fruit, while in the other house, the fruit was so badly stained that every tomato had to be wiped although the number of insects present was visibly less. If the vegetable lacks something which the caterpillar or insect needs, he may go on eating long after satiety in an effort to find it. As already stated, this is a purely personal hypothesis. I have no proof that it is so, it merely seems to me probable in view of the evidence available.

III

MAN

If it should be conclusively proved that humus, whether through the agency of soil fungi or otherwise, can restore to plants some virtue otherwise missing, something which can best be summed up in the word 'quality', embracing as it would seem, palatability, keeping quality and health, and that such plants do act as a conveying agency, passing on this quality to animals; then the foundation would be well and truly laid for further investigations as to the extent to which this same quality can be further transmitted to the human race. The claim that it can be so transmitted has already been made, but the moment we attempt to provide evidence involving human beings, we are faced with a much more difficult task, critical proof is almost impossible to obtain, for you cannot treat people as though they were laboratory guinea-pigs. Nevertheless, many indications supporting the claim *do* exist, and a few examples must be given.

In many districts of India rice is the staple diet. I have been told that in those areas where the rice crop is sown broadcast in the paddy fields, the standard of health of both crop and population is low, but that in other districts where it is the custom to sow the rice in a seed-bed of soil rich in humus, and thence to transplant the seedlings, both crop and people are remarkably healthy.

That the incidence of malaria may also be connected with methods of crop cultivation is indicated in the following quotation.

'There are multitudinous examples of the retreat of the crop and of the animal and of mankind before the parasite, but we are only now beginning to get examples of the reverse process. In two cases—malaria and sleeping sickness—there are signs that soil fertility is the real method of dealing with these diseases.

'I will take malaria first. In one of the most intensely malarious areas of India, the Terai—a strip of forest at the base of the Himalayas—malaria is so bad in the early rains, June to September, that it is regarded as a death trap both by Europeans and Indians. Nevertheless, in this area a tribe exists who are practically, to all intents and purposes, immune to malaria. They go in for intensive agriculture and their villages are very clean. . . . What we want is a McCarrison to follow this clue out still further. That malaria depends on the way crops are grown is supported by other facts. In Western Bengal where rice is not properly grown, there is intense malaria; in Eastern Bengal (Mymensingh) where rice is exceedingly well grown, there is very little malaria. Further, in the rice areas, when these are invaded by the water hyacinth, the conditions for malarial mosquitos seem to be removed. Incidentally, the water hyacinth will provide the humus needed to grow rice really properly.

'As regards sleeping sickness, the evidence is not so complete: nevertheless it exists. In Nigeria it has been found that the use of cattle manure for raising fodder crops is followed by a distinct increase in resistance to the Tsetse fly disease. There are some indications that the same thing occurs in Tanganyika. What we really need in the fly belt of Africa is a fairly large area of really fertile soil, so that we can see what the effect of this is on the incidence of sleeping sickness.'[1]

In East Africa 'Major Layzell found that the vegetables grown for his labour force on the land manured with humus, made largely from sisal waste, resulted in a marked improvement in the general health, physique and efficiency of his workers. The men performed their tasks much more easily than was the rule before the new system of nutrition was introduced. Major Layzell is now engaged in starting this work at a new centre.'[2]

Corroboratory evidence comes from a large preparatory school near London at which both day boys and boarders are educated. The vegetables consumed in the school are provided from its own garden. Until a few years ago these were raised with artificials, then a change-over was made to Indore compost. The headmaster records the result as follows:

'Formerly, in the days when artificials were used, cases of colds, measles, and scarlet fever used to run through the school. Now they tend to be confined to the single case imported from outside. Further, the taste and quality of the vegetables have definitely improved since they were raised with compost.'[3]

This experiment is of particular interest when it is considered in conjunction with a somewhat similar one undertaken in New Zealand.

[1] *Compost News Letter No. 2.* (Howard quoting Major Clyde's report.)

[2] Howard, letter to *The Times*, 10 August 1939.

[3] Personal letter to Sir Albert Howard from the headmaster concerned. [This school is now closed, but in 1953 I was able to get first-hand details from two co-educational boarding schools in the USA with an exactly similar story to tell. One was in California and the other in upper New York State. As far as I know, the latter is still in being. For a description of these see *Mother Earth*, January 1954, or *9,600 miles in a Station Wagon* by Eve Balfour, both obtainable from the Soil Association—for address see p. 370.]

The truly devastating effects of soil exploitation, past and present, are nowhere more evident than in that beautiful Dominion.

Mrs. Ysabel Daldy, founder of the Physical and Mental Welfare Society of New Zealand Incorporated, has been kind enough to supply me with the official facts, figures, and statistics which fully endorse this statement. 'With a perseverance worthy of a better cause,' she writes in a summary of the situation, 'New Zealand has for years past been carrying out a nation-wide experiment whose outcome has proved beyond reasonable doubt that a people reared upon eroded and otherwise exhausted soils becomes a people whose condition gradually deteriorates.' She quotes Professor Worley, MA, DSc (Chief of the Department of Chemistry, University of New Zealand) as stating that when New Zealand was first colonized the country was 'almost entirely covered with rich vegetation, chiefly forest, which covered the hills and extended up the mountain sides to a height of some 3,000 to 4,000 feet. The floor of this forest was a porous layer of priceless humus, the product of thousands of years of formation. In this humus were the mineral salts extracted from the subsoil and the rocks, by the roots of the plants and trees. In our unthinking exploitation of the land we have destroyed the forests over the major portion of the country and millions of pounds' worth of soil fertility have been sent sliding into the sea by erosion—one of the consequences of deforestation.'

'The whole of New Zealand food supplies,' writes Mrs. Daldy, 'are now grown from soils manured with ordinary chemical fertilizers,' and gives Professor Worley's comment on this practice.

'By what we add, as well as by what we fail to restore to the soil, we are profoundly affecting its chemical composition, its biological content and its physical nature. We are thus affecting the quality of the food grown on such soil, and, in consequence, the health and vitality of the population. It is now recognized that much of our food has serious deficiencies, and that very many of our ills are due to this cause.'

What are these ills? It is said that figures speak; in New Zealand they shout. During the twenty years between the two world wars New Zealand's population (by births and immigration) increased 36 per cent. In the same period the birthrate decreased from 23 to 17 per thousand of the mean population; mental cases increased by 100 per cent; and admissions into public hospitals showed a gradual increase up to 126 per cent. Maternal mortality is a little higher in New Zealand than in Great Britain. Infant mortality is certainly the lowest in the world, but the incidence of illness among New Zealand toddlers between two and five years of age is alarming. Every year, of children of pre-school age, some 80 per cent are found to be physically defective in some way.

Of every 100 children who enter New Zealand schools fifteen show signs of needing medical attention, fifteen need observation, many show signs of nose and throat trouble, and at least two-thirds have dental caries. In this connection, the New Zealand Ministry of Health has published the fact that 30 per cent of all pre-school children suffer from nose and throat troubles, 23 per cent suffer from gland troubles, and 2 per cent have some form of lung trouble. The official figures for illnesses among children at school are: 5 per cent suffering from enlarged glands; 15 per cent suffering from incipient goitre; 15 per cent suffering from enlarged tonsils; 32 per cent suffering from dental caries; and 66 per cent suffering from other physical defects.

Lord Bledisloe reports that during his term of office as Governor-General of New Zealand he ascertained that about 60 per cent of the inmates of the Dominion's public hospitals had previously suffered from malnutrition. He has also stated that 'for many years natural soil deficiencies have caused widespread animal disease, bush-sickness being found traceable to a lack of iron and cobalt, dopiness in sheep to a lack of lime, Waihi disease to a lack of phosphates, and goitre to a lack of iodine.'

Mr. H. B. Tennent, the agricultural editor of the *Weekly News* (New Zealand's leading agricultural journal), wrote in 1938: 'Of all primary producing countries in the world, New Zealand is probably most lacking in some important soil elements essential to healthy plant and animal growth. Animal sicknesses and the rapidly increasing populations of hospitals and mental institutions can in a great many cases be directly traced to mineral and other deficiencies in the foods produced from improperly balanced soils.'

It was against this background that Dr. G. B. Chapman[1] carried out a notable feeding experiment. An article by Mrs. Daldy describing it, appeared in the issue of *Nature* for 8 June 1940, and I am indebted to the editor of that journal, as well as to the author, for permission to quote from it. It was begun in 1936 at the hostel of the Mount Albert Grammar School, the second largest grammar school in the city of Auckland. The subjects were 'some sixty boys, teachers and staff'.

'At the time of the inception of the experiment,' writes Mrs. Daldy,

[1] Of the Physical and Mental Welfare Society of NZ Inc. Now the President of the NZ Humus Society.

'the dietary at the hostel was liberal, being well above the customary standard for boarding-schools; yet the boys consistently suffered (as was the case in other New Zealand schools and institutions) from colds, catarrh, septic tonsils, epidemics of influenza, dental caries, and other preventable complaints. . . .

'Dr. Chapman opened his campaign in 1936 by delivering a few short lectures to the resident teachers and boys, advising the growing of the hostel's fruit and vegetables from soils to be treated by properly prepared humus. He was successful in arousing the interest of the teachers, the boys, the matron and the staff. The reform was put in hand and the change made from "chemically grown" fruit, salads and vegetables to the "naturally produced" foods now in use.'

Since much of the boys' food, such as meat, bread, etc., had still to be obtained from outside sources, this portion of their diet continued to be the produce of 'depleted soils'. Dr. Chapman therefore gave the boys a daily extract of vitamins A and D to counteract the deficiencies in these foods. The Matron of the school has submitted a report on the results of the experiment.

'The first thing to be noted,' she wrote, 'during the twelve months following the change-over to garden produce grown from our humus-treated soil, was the declining catarrhal condition among the boys. Catarrh had previously been general and, in some cases very bad among the boys. In specific cases the elimination was complete. There was also a very marked decline in colds and influenza. Colds are now rare and any cases of influenza very mild. Coming to the 1938 measles epidemic, which was universal in New Zealand, the new boys suffered the more acute form of attack: while the boys who had been at the hostel for a year or more sustained the milder attacks, with a much more rapid convalescence.

'During the past three years there has been a marked physical growth and development during terms of heavy school work and sport' (actual heights and weights are quoted). 'In some cases boys go through a period of indisposition for several weeks after entering the hostel. This would appear to indicate that the method of feeding causes a certain detoxication period which when cleared up does not return. Excellent health gradually ensues in all cases, and is maintained. There are fewer accidents, particularly in the football season, which would possibly indicate that the foods in use contain the optimum amount of minerals and vitamins, thus ensuring a full development of bone and muscle and a greater resiliency to fracture and sprains. The satisfactory physical condition described is maintained during periods of rapid growth and development of mind and body. Constipation and bilious attacks are rare. Skins are clear and healthy, while the boys are unceasingly active and virile.

'Since the change to naturally grown garden produce, the periodical reports in regard to the boys' dental condition have been more than gratifying.'[1]

It may be argued that the remarkable improvement in health brought about by this experiment was due to the vitamin treatment and not to the altered method of vegetable culture. Obviously further experiments are required to provide definite proof on this point, but when considered in conjunction with other evidence presented in this chapter, particularly the cases immediately preceding and following this New Zealand example, and when due weight has been given to the proved deficiency in so much of New Zealand's soil, then it must be conceded, I think, that Dr. Chapman's experiment serves considerably to strengthen the indications.

My next example is also of more than usual interest, for it comes from Dr. J. W. Scharff, who until the tragic fall of Malaya was Chief Health Officer at Singapore. Here is his firsthand account as published in *Mother Earth*, the journal of the Soil Association.[2]

'The editor of this magazine has kindly invited me to record the sequel to my wartime effort of introducing the Indore method of composting into the village life of certain communities in Malaya.

'The interruption of these developments following the Japanese invasion is so complete that the picture now presented is almost the reverse of what was then in progress. It therefore remains for me to describe the story of the experiment and to amplify some of the reasons which lead me to the conclusion that there must be exceptional quality in fresh compost-grown foodstuff. I believe that the further development and understanding of this subject is of the utmost importance.

'The original intention of the experiment had been to prepare Singapore and neighbouring territories against the advent of war by increasing food production. It was evident that if organic wastes could be fully utilized in compost-making, the danger of starvation in the event of a blockade around the coast of Malaya might be reduced.

'From January 1940 until January 1942 I therefore had a unique

[1] At the beginning of the war 40 per cent of the recruits volunteering for New Zealand's armed forces were temporarily rejected because of the condition of their teeth.

[2] See p. 370.

opportunity, due to wartime needs, of watching the progress of a campaign for growing vegetables, and seeing that they were eaten by a labour force of nearly 500 Tamil coolies. These men were employed by the Singapore Health Department in various parts of the island of Singapore.

'As soon as England became involved in war this emergency made it possible to allocate an area totalling about forty acres of vegetable allotments on favourable terms to men engaged in sanitary duties. My labourers were granted these allotments on condition that they prepared compost and used the vegetables and fruit grown therein for themselves and their families only. Sale of the produce was not allowed. Thus it was ensured that these goods were used at home. The local Agricultural Department lent their inspectors and staff to teach the men how best to grow vegetables, and demonstrations in cooking and preparation of the foodstuff were organized for each of the labour settlements. Compost-making was started on a large scale and during the months previous to opening of the campaign a supply of over a thousand tons of compost was ready to launch this experiment.

'During the course of the ensuing months, apathy and indifference on the part of the labourers gave way to interest and enthusiasm as soon as it became apparent how well plants would grow on soil rendered fertile with compost. A number of vegetable shows were arranged, at which the healthy produce of fertile soil was exhibited and prizes were awarded. Within six months the accumulated stocks of compost were used up and more active steps were taken to augment the supply as well as to satisfy the growing demands of other enthusiastic gardeners inspired by the achievements of my men.

'At the end of the first year, it was obvious that the potent stimulus to this endeavour was the surprising improvement in stamina and health acquired by those taking part in this cultivation. Debility and sickness had been swept away, and my men were capable of, and gladly responded to, the heavier work demanded by the increasing stress of war. But for the onslaught by the Japanese which overwhelmed Malaya, I should have been able to present a statistical record of the benefit resulting from this widespread effort of vegetable culture on compost such as would astonish the scientific world. The results were all the more dramatic in that I had not expected this achievement.

'The numbers taking part in this venture were so large as to preclude any possibility of mistake. It might be argued that the improvement in stamina and health amongst my employees was due to the good effect of unaccustomed exercise or to the increased amount of vegetables consumed. Neither of these explanations would suffice to explain the extent of the health benefit observed amongst the women, children and dependants of my labourers who shared in this remarkable improvement.

'Shortly before the tragic disaster which brought Singapore within the hateful grasp of the Japanese invader, it became apparent that the health of men, women and children, who had been served consistently with healthy food grown on fertile soil, was outstandingly better than it was amongst those similarly placed but not enjoying the benefits of such health-yielding produce.

'The scene on my return to Malaya with our victorious army early in September 1945 was vastly different from that which was observed during the previous period of plenty and prosperity. Most of the labourers employed in scavenging, and their families, had been obliged to leave their homes to seek other work. Many were enslaved by the Japanese for enforced labour on the railways in Siam. Soil cultivation was neglected and there was no security for those who attempted to carry on. Produce was stolen and equipment destroyed; misery, sickness and starvation stalked throughout the land.

'This is the broad outline of the passage of events up to the time of our triumphal re-entry into Singapore less than a year ago. Since then, thanks to the energy and enthusiasm of my colleagues, a great deal has been done to repair the devastation and to revive the organization which yielded such satisfactory results. Still many more years of toil and effort will be needed to approach the point at which the foregoing experiment on soil cultivation can be repeated in Malaya. In the meantime, there is urgent need of better knowledge of the underlying principles in the health benefits coming from fresh food grown in compost-fed soil.

'Mr. Donald Hopkins[1] has recently questioned my conclusion that these good effects were attributable to the compost, and suggests that vegetable produce grown in any other way might have given equivalent results. Such criticism is welcome if it leads to further investigation on compost-grown foodstuff, which the Soil Association so urgently desires; so it may be useful to take this opportunity to point out that my observation was made upon a stable labour force, consisting of men who had been in regular employment for many years.

[1] Donald P. Hopkins, *Chemicals, Humus and the Soil*, Faber & Faber, London, 1946.

Malnutrition had not previously been apparent amongst them, and there was no shortage of ordinary vegetable produce in the village markets. The women and children were in regular attendance at rural welfare centres, and the men were in receipt of a regular salary amply sufficient to satisfy their needs. The reason for my insistence in getting them to grow and eat compost-grown vegetables was to promote a system of sanitation which would make the community, as nearly as possible, self-supporting when the tide of war spread against Malaya. The healthy response which eventually greeted this undertaking was unexpected, and was one which I had not hitherto experienced in many years of trial on the means of improving public health. The conclusion that "an oasis of good health had become established founded upon a diet of compost-grown food" therefore rests upon a firm foundation. At the present time, when shortage of food supply is a major issue, this lesson of Malaya's wartime experience is one which should be considered and confirmed by further studies.'

I hesitate to advance my own personal experience among these examples, but I have decided to do so for two reasons. Firstly, because I think it merits the appellation of an 'indication', and secondly, because people who recommend any course of human action or behaviour which lays itself open to the accusation of claiming to be a 'cure-all', so often appear to others to be themselves more than usually in need of the 'cures' they advocate. I feel, therefore, that I owe it to my readers to forestall the question, 'Have you tried it on yourself?'

I have lived a healthy country existence practically all my life, and since 1918 I have been actively engaged in farming. I am physically robust, and have never suffered a major illness, but until 1938 I was seldom free from some form of rheumatism, and from November to April I invariably suffered from a continual succession of head colds, while my annual visit to the dentist always involved a series of 'stoppings' and occasionally an extraction.

In 1938 I was introduced to the humus theory, the McCarrison experiments, and the effect on health of 'whole' diets as described by Dr. Wrench in *The Wheel of Health.*

In so far as I was able I immediately put the theory into practice. I started to make compost by Howard's method using it first on the vegetables for home consumption. After harvest I saved some of my wheat crop from a field which for several seasons had received only farmyard manure. This I ground, just as it was, on the ordinary farm mill kept for grinding grain for livestock. Thereafter, in place of the baker's loaf, I ate home-made bread baked from this home-grown, home-milled, whole wheat.

That winter, 1938–9, I had no colds at all, and almost for the first time in my life was free from rheumatic pains even in prolonged spells of wet weather.

In 1939 the only wheat available was grown on rather poor soil deficient in humus, and that winter, 1939–40, much to my annoyance and disappointment I had a slight return of winter rheumatism and caught three colds. But it was the first war winter. Rationing and other restrictions had been introduced, and thus there were other differences in my diet. Notably, less butter (for we did not make it) and no cream, previously plentiful. I therefore thought it probable that if my partial relapse was due to diet at all, any, or all of these differences might be responsible.

In 1940 I was once more able to save wheat for home baking that had been grown on fertile soil. Restriction in other portions of my diet of course remained. That winter, 1940–1, my rheumatism, except for one short attack lasting about a week, again disappeared, and I caught one cold only, and this despite exceptionally long working hours by day, and nights constantly disturbed by civil defence duties. Up to the time of publication (autumn 1943), this state of good health has continued. I have had no colds or rheumatism this winter.

Now as to teeth. I had my usual crop of stoppings in the summer of 1938. A year later, during the autumn of 1939, for the first time that I can remember my dentist could find nothing to do. I did not visit him again until October 1942. All that then needed to be done, after this interval of four years, was one small stopping in a place where an old filling had worn away.[1]

I do not claim for this experience that it provides anything in the nature of proof (though it certainly provides me with every inducement to continue to eat compost-grown food). The interest lies in the fact that this is a case in which no change took place in the *articles* of diet, only in the way in which they were grown.

[[1] Since I *can* follow up my own case, I feel I should. My record of good general and dental health continued until 1959. That year I was away from home for twelve months, touring Australia and New Zealand for the Soil Association. My teeth suffered from this, but the amount of attention required on my return, and since, has been minimal compared to that usually required at my age. Since I retired in 1965 (in spite, oddly enough, of this being from a heavy to a light soil) I cannot claim to have been entirely free from a mild form of rheumatism in damp weather, but I still work my compost kitchen garden single handed, and am able to keep our small household self-supporting in vegetables and fruit (other than citrus, of course) so at seventy-six I am not doing too badly!]

Lastly, I conclude this list of indications by quoting three cases experienced by members of the Panel Committee of the Doctors of Cheshire, as reported in the references to the 'Medical Testament'. These cases are of great interest, especially the first, which is I think the most complete account of a whole diet experiment yet to be recorded in this country.

EXPERIENCES OF MEMBERS OF THE COMMITTEE

'*These have no pretensions to be scientific research but are submitted as examples of the effects of diet noted over and over again in ordinary family practice.*

A

'The reality of the value of using fresh, well-chosen food is shown in a practice in a Cheshire village. The County antenatal scheme makes provision for a woman's own family doctor to supervise her in pregnancy. Her nutrition is his first concern. In the village referred to the local mothers' organization conducts a "child-welfare" each month to which the local doctor is honorary MO. A fair percentage of the mothers of the neighbourhood, mostly, but not all, his patients, attend and thus receive his advice upon the nurture of the children some eight to twelve times a year; and a thorough mutual understanding has grown up. The food of the mother, during her pregnancy, is wholemeal bread, one to two pints of milk (raw), generally including half a pint at breakfast taken with porridge (medium oatmeal scattered into boiling water and stirred till it thickens); eggs are used freely; salads in abundance, including celery and dandelion leaves; green leaf vegetables plunged into boiling water for five minutes and eaten with butter or poached eggs—or with meat, but the amount of meat taken is very moderate; liver weekly; herrings twice or once a week and a little cod liver oil except on herring days; fruit in abundance—such is an outline of their food. The evening meal is often begun with soup of the Scots broth type, but carrots, unpeeled, are grated into it just before serving. Apples are eaten in their skins or baked in their skins, and other fruit is used freely. Potatoes are baked in their skins or boiled in their skins (dropped into boiling water and boiled till they "smile", then the water is poured off and a crumpled cloth put in the pot whilst it is drawn to the side of the fire). Cheshire cheese, grated into a salad, with a hard egg, is advised and popular.

'The wholemeal bread in question is fertility bread, that is, locally grown wheat, ground (or rather dashed to pieces by a steel fan revolving 2,500 times a minute) in a local mill, mixed with half its weight of raw wheat germ fresh off the rollers of a Liverpool mill and —a point to be rigidly insisted upon—baked at once, within thirty-six hours at most—a rather close but very palatable bread, requiring no little skill in baking, though a number of bakers in the neighbourhood have acquired it.

'If her haemoglobin be 80 per cent or under, the mother receives iron, ferrous sulphate, or other.

'With rare exceptions, and those almost always "strangers", the mothers feed their infants at the breast nine months and then wean them by a year or a little more. The nursing mother's food continues as in pregnancy, including the greens and salads. The feeding times are 6 a.m., 10 a.m., 2 p.m., 6 p.m. and 10 p.m.—and very seldom in the small hours.

'The children begin to bite the wholemeal crusts at from seven to nine months and then often get a little raw turnip juice made by putting Barbados sugar (Muscovado) into a hollowed-out swede or other turnip.

'Except custard (egg and also "beast milk") and junket, no milk puddings are used, except when occasionally rice which still has its germ and silver skin, can be obtained. Ordinary rice is discouraged.

'Furmity (Frumenty) is not yet forgotten by the indigenous people and its use is encouraged.

'No patent or "processed" food of any kind whatever is employed, with the exception of Marmite and of dried yeast.

'Broth, red gravy, brains and marrow bones are used for the children from nine months onwards.

'The children are encouraged to go barefoot, which suits them well, but only a very few of the mothers, so far, are entirely cordial about this.

'When the regime grew up little by little, many years ago, there were a thousand minor difficulties; fruition was slow. But it has now been established long enough for the generation it has fostered to be studied.

'The teeth are a good index of the fidelity of the mother in carrying out the regimen, both before and after the baby was born; and it may be said that perfect sets are becoming more common.

'The children are splendid. As infants they sleep as well as could be wished, grow well, are not over fat but weigh well and very seldom "ail anything". Broncho-pneumonia, for instance, is almost unknown amongst them. One of their most striking features is their good humour and happiness. They are sturdy-limbed, beautiful-skinned, normal children.

'It is not desired to give the impression that the child population of this village is perfect or that complete compliance with the dietary advised is secured even amongst all who attend the centre; but it is a fact that the mothers follow it substantially and with good results, which those concerned think they recognize.

'The benefits are visible in the households. "We have all taken to brown bread now; and I'm sure we're the better for it."—"No white bread comes in this house: it's all wholemeal and there's no trouble with constipation."—"We've got that fond of the meal bread that when we went to the seaside and they gave us white we all looked at each other!"—And a number of families seriously cultivate their gardens for growing a succession of saladings.[1]

B

'A young woman, town bred, resident in a large public institution where she was a technical instructress, married a farmer. That is to say she passed from a *ménage* where white bread and contract butter were the invariable rule whilst green vegetables were the rare exception and subjected to prolonged cooking in steam jacketed pans at that, to one where good fresh food was available if trouble were taken. It was taken; but it was too late. Anaemic and constipated as she was, her pregnancy confirmed the fear that she was in no fit state to become a mother. Despite all efforts she passed through the kind of experience which seems to justify the notion that to be gravid is to be ill. All forms of the toxaemias of pregnancy, anorexia, hyperemesis, albuminuria and oedema were exhibited by turns; and a foul otorrhœa of old standing awoke to virulent activity. Despite all this she arrived at term and underwent an anxious, lingering and difficult confinement. The infant, resuscitated by prolonged artificial respiration, was sickly and emaciated. His respirations were intermittent and after some sixty hours he breathed his last. The mother was febrile and exhausted, her condition anxious; but she hung on to life and almost

[1] Proper compost making is fostered in the cottage gardens. Instruction in how to make it is given, and garden competitions are held for the best vegetables grown with compost (author).

imperceptibly her condition drifted—or rather was led, for she was surrounded by the utmost solicitude—into a slow convalescence.

'All this detail has been related as a contrast to the sequel. Being active-minded she began to take a quickened interest in her surroundings, the farm. Little by little she came to do a real share of the normal work of a country-bred farmer's wife. Strength and the look of health came to her. The improvement was not fortuitous. It was decided from the start that a real attempt should be made to rebuild her physique. Her food was the fertility bread, farm or Empire butter, salads in variety and abundance, with grated carrots, soup and barley, potatoes in their jackets and unlimited fruit. Marmite and dried yeast were used. She took about a quart of milk a day fresh from their own cows, and she gradually came to take as many eggs as she wished, herrings freely and occasionally liver. She enjoyed life and it was obvious in about a year that she was in abounding health.

'Then she announced that she was again pregnant. A time of excellent health followed, no drawback. The confinement was normal, birth being spontaneous. The baby was in first-rate condition, lactation was established and maintained fully and easily for over nine months, finishing finally in about a year. With the exception of a mild influenzal attack, the mother's health has been good. Furthermore, the chronic otorrhœa of many years' standing has at last dried up.

'The baby is as good physically and in morale, as could be wished.

C

'In May 1938 I was consulted by a young Irishman. His age was about twenty-three. He complained that he was "very poorly and had been sick and vomiting for three days". I found he was suffering from catarrhal jaundice. He had come over from the west of Ireland two months previously and was employed on a road construction scheme in Cheshire.

'I questioned him about his diet. I found that his breakfast consisted of bacon, white bread and tea. His dinner he took to his work and was mainly sandwiches of white bread and ham or beef with tea, and his evening meal was white bread and butter, sometimes an egg, and sometimes a bit of meat and tea.

'My examination revealed to me, despite his illness, a physique and an alertness of mind and body which it was a delight to behold. He was well over six feet in height with jet black hair and a ruddy complexion, and in his jaws there were thirty-two healthy and symmetrical teeth. He possessed a supple body and limbs of good proportion, and apart from his gastric condition was of sound constitution.

'I thought him to be as fine a specimen of humanity as I had ever seen. He aroused my interest and I wondered how had such a body been nurtured.

'On inquiry, he informed me "they were very poor in the west of Ireland and food was very poor indeed". "What did they have for breakfast?"—"Porridge and milk and a little bit of fat bacon."

' "What did they have for dinner?"—"Porridge and milk and a little bit of fat bacon, buttermilk and potatoes, and sometimes broth." His mother made grand broth of vegetables which they grew in the garden—carrots, turnips, potatoes, green vegetables.

' "What did they have for tea?"—"Oatcakes, scones and butter with milk and plenty of buttermilk, sometimes a little tea with bread and eggs." On further inquiry I found they might have an old fowl with bacon and potatoes for their Sunday dinner. "His mother made very good gooseberry and blackberry jam."

' "Did they have any fish?"—"Oh yes, they always had a bit of dry salted fish and as they lived near the river they sometimes got a bit of salmon." "They had very little meat, but they could always get a rabbit." "The potatoes were often baked for supper."

'It would appear that his body had been nurtured on the natural products of the west of Ireland.'

What happens when complete communities live for successive generations on whole diets raised on fertile soil will be described in the next chapter.

Note

The main argument developed in this chapter is that disease resistance, in so far as it is due to the immunity of the host, depends on the ability of correctly nourished tissue to prevent the successful attack of pathogenic organisms. It is fully realized, however, that immunity may also follow when the medium is rendered unfavourable to the growth of pathogens owing to the activities of other organisms, whether by competition for nutrients, or by the production of inimical by-products. Recent researches on certain moulds—e.g. species of *Penicillium*—suggest the possibility that factors responsible for the increased disease resistance observed in compost-grown plants may be connected, in the broadest sense, with the ecological interrelationships and activities of members of the soil population. At present one can only surmise that such factors may also be operative in the animal kingdom.

Nevertheless, as each year passes and more discoveries are made about living processes, evidence in support of the ecological view accumulates, and particularly are we being made increasingly aware that the study of nutrition is a very complex one indeed.

In the course of the 1950 Sanderson-Wells Lecture, which he delivered at the University of London, Colonel Sir C. Stanton Hicks, MSc, PhD, MD, FRIC, said:

'Until the recent war, the body was pictured as a machine, the structure of which, once completely developed, would, if wear and tear were replaced, and vitamins thrown in as a sort of anti-knock remedy, remain in going order.

'Today, the whole concept has been swept away by the application of isotopes in metabolic research. Instead of a machine that is built and maintained in repair, the body, seen through the eyes of a new observational procedure, is an appearance resulting from equal rates of comings and goings of the material substance that is our food. As a whirlpool remains, whilst the stream flows, so does the living body of the organism remain, as long as the flow of material substance is appropriately maintained.

'This new outlook quite alters our basic notion of nutrition, for it becomes easier to see that health depends on the material nature of the stream of nutrients. Concerning the quantitative and qualitative nature of this stream we have but the vaguest notion, and vitamins and trace elements begin to appear as a vitally important key to the dynamic interchange within the cell itself.

'I have always taught that the discovery of the vitamins was in the nature of a warning that food is a much more complex matter than we have ever thought, and that, individually, vitamins were but another scientifically interesting observational item.

'If we proceed to the soil whence comes this stream of being, we find ever-increasing evidence of biological complexity, where chemical simplicity has for so long been postulated and made the basis for action. It is no more true today to assume that the soil maintains plant and animal life by being an inert store of chemical elements, than it is true that food is merely a source of energy and repair material for a completed adult body structure.'

Technical Reference

A

PREDACIOUS FUNGI

(Dr. C. Drechsler, *Biological Reviews*, XVI, 1941)

'When Rouband and Descazeaux (1939) added larvae of strongyloid nematodes parasitic on the horse, belonging to the genera *Strongylus* and *Tribonema*, to agar cultures of *Arthrobotrys oligospora* and *Dactylella bembicodes*, the very active strongyloid specimens were captured and consumed in quantity. *D. ellipsospora* showed similar though lesser predacious capabilities in relation to these parasites. However, the very sluggish rhabditiform larvae apparently evoked very little development of predacious apparatus.

'Descazeaux (1939a), in determining that strongyloid larvae of the genera *Strongylus* and *Tribonema* were captured by *Arthrobotrys oligospora* and *Dactylella bembicodes*, made use of nematodes newly hatched from a pasty mixture of dung and powdered charcoal. The mixture was kept confined in crystallizing dishes covered with inverted bases of Petri-dish cultures of the two fungi. Several strips of moist paper placed vertically in the containers provided passageways suitable for migration of newly hatched animals from the dung paste to the agar cultures. Predacious organs, the formation of which began 24–36 hours after the relatively early arrival of rhabditiform larvae, operated especially effectively against the tardier strongyloid larvae, with the result that in 20 days all the larvae that emerged from the dung were exterminated. In a later paper Descazeaux (1939b) reported that larvae of the family Trichostrongylidae, which are parasitic in the digestive tracts of cattle and sheep, were captured and killed in predacious apparatus formed after their addition to cultures of *Arthrobotrys oligospora* and *Dactylella bembicodes*. Likewise, when conidia of either fungus were added to water containing such larvae, predacious organs were formed by means of which the animals were captured and killed.'

B

'Deschiens and his colleagues admit that possibly even under ordinary natural conditions the predacious hyphomycetes in some measure carry on destruction of infectious nematode larvae. By providing for increased mycelial development through application of fungus material—the increased development reported by them seems greatly in excess of expectations—they aim to augment the activity of the predacious hyphomycetes to the point where it becomes effective as an important subsidiary, if not principal, prophylactic factor. Manifestly their treatment by extensive application of fungus material departs markedly in principle from the treatment investigated in Hawaii, whereby large quantities of organic matter are incorporated in the soil. The former treatment would seem to be based on the idea that nematode-capturing fungi are meagrely distributed in nature, and yet are addicted to very aggressive saprophytic development when brought into contact with foul natural substrata; whereas the latter treatment postulates rather that the fungi in question are very generally distributed in soil and decaying materials, that they subsist habitually on nematodes, and that accordingly they can best be stimulated in vegetative development and useful predacious activity by first supplying them with larger quantities of saprophilous living prey.'

7

Whole Diets

[It is very fortunate that the examples of healthy communities cited in this chapter were so well documented, because they paint a picture that is not wholly true today. Modern transport has brought these peoples in contact with Western culture and food habits (refined carbohydrates and packaged goods) which have led to deterioration of their original outstanding physical fitness. Some isolated communities, virtually without physical or mental defects, do still exist. The best book on this whole subject is the late Dr. Weston Price's *Nutrition and Physical Degeneration* published by Price & Pottenger Foundation, obtainable from 'Whole Food', 112 Baker Street, London, and admirably illustrated. I can thoroughly recommend it to specialists and laymen alike, but it is a must for students of nutrition.]

In the Introduction to his book, *The Wheel of Health*,[1] Dr. Wrench makes this thought-provoking statement: 'After debating the question —Why disease? Why not health?—again and again with my fellow students, I slowly, before I qualified, came to a further question—Why was it that as students we were always presented with sick or convalescent people for our teaching and never with the ultra healthy? Why were we only taught disease: why was it presumed that we knew all about health in its fullness? The teaching was wholly one-sided. Moreover, the basis of our teaching upon disease was pathology, namely, the appearance of that which is dead from disease.'

This view, that the professional attitude to sickness is one-sided is shared by the compilers of the PEP Report (1936) on the British Health Services. The authors express it as follows:

'Health means more than not being ill. A new attitude is needed, involving not so much a departure from the old as a more thorough grasp of the different elements in health policy. Many people are at any given moment suffering from defects, injuries or sickness so pronounced as to make them unable to carry on ordinary occupations

[1] Unless otherwise stated, all quotations in this chapter are taken from this book.

and leisure activities. These are the "cases" with which a large part of the organized health services mainly deal. But in addition there are far larger numbers of people suffering temporarily or permanently from less acute defects, injuries, or inadequacies, which are not sufficient to unfit them for work or play, and may not even be noticed at all, but nevertheless suffice to place them in an unnecessarily weak position for creating and maintaining good physique, energy, happiness, or resistance to disease. . . . No contemporary health policy can be considered adequate which does not deal with the second group as well as the first. . . .

'While efforts at effecting the cure of diseases cannot be relaxed, efforts at prevention of ill health can and must be increased. The aspect of raising standards of nutrition and of fitness should be given much prominence. Health must come first: the mere state of not being ill must be recognized as an unacceptable substitute, too often tolerated or even regarded as normal. We must, moreover, face the fact that while immense study has been lavished on disease no one has intensively studied and analysed health,[1] and our ignorance of the subject is still so deep that we can hardly claim scientifically to know what health is.'

The theory which I have endeavoured to expound in this book is that the only true conception of health is one of wholeness, dependent upon both the continuity and the completeness of the cycle of life. I shall make no attempt to discuss the philosophical aspect of this conception, I am concerned only with presenting certain evidence suggesting that it is biologically sound. For the sake of clarity in this presentation, the argument has been divided into two parts. The first states that the determining factor in health is food, and the second suggests that the health-giving property of food is dependent on the way it is grown, prepared and consumed.

There is always an inherent danger in making arbitrary divisions where no true division exists, but in the present case this could hardly be avoided. The risk of failing to see the wood for the trees must sometimes be taken in order to discover how large a number of trees go to make a wood. In the previous chapters—to pursue the same analogy—I have invited you to examine some of the trees. In this

[[1] When Dr. Wrench wrote this, he was clearly unaware of the intensive study of health carried out by Drs. G. Scott-Williamson and Innes Pearse known as 'The Peckham Experiment', and the definition of health that resulted from it. (See Part II, also new introduction to Part I.) So far, however, this extremely important and inspired piece of research remains the only investigation of its kind ever to have been undertaken.]

chapter I want you to take a look at the wood as a whole. This, research has, so far, largely failed to do. Research workers in chemistry, biology, mycology, botany, veterinary science, and medicine, have for too long been working in watertight compartments, each busily dissecting his own tree, until in the process the wood has become so sadly dismembered it is small wonder that we sometimes cease to be aware of its existence.

It may be objected that I have followed their example of fragmentation in laying so much stress on the role of the fungus in nutrition. In so far as this accusation is well founded, my defence is that this is the link in the life cycle which is most frequently omitted, but which is at the same time the easiest to restore. In emphasizing its importance, however, I have never intended to suggest that it forms other than a part, however important in itself, of the complete cycle of nutrition and health which is wholeness.

McCarrison in the Cantor Lectures states: 'The diet of the Sikhs is only health-promoting so long as it is consumed in its entirety', and as Dr. Wrench points out, with their whole diet these people 'have preserved the wholeness of their health, a thing which we have failed to do'.

'In the writings of the scientific experts on nutrition, there are very numerous part-diet experiments based on synthetic or specially made-up diets, omitting or cutting down the quantity of one or more of the factors which compose a diet. One scientist will cut down the quantity of protein given and watch the effect of this upon animals; another will cut down the fats and note the resulting sicknesses; another will give vegetable or irradiated vegetable fats in places of customary animal fats; another will give a diet in which vitamin A is defective, B is defective, C is defective, and so on.

'The experiments are skilfully devised and carried out with consumate technique. They lead to a mass of knowledge about proteins as things in themselves; fats as things in themselves; vitamins as things in themselves; but whether these can be things in themselves and are not really relative to a host of other conditions in nutrition is as yet scarcely considered. . . .

'Our health or wholeness has fragmented no less than our diet. A swarm of specialists have with the invention of science settled on the fragments to study them. A great deal is found out about each several disease; there is a huge, unmanageable accumulation of knowledge, and this and that disease is checked or overcome. But our wholeness has not been restored to us. On the contrary, it is fragmented into a great number of diseases and still more ailments. We have lost wholeness, and we have got in its place its fragmentation with a multiplexity of methods, officially blessed and otherwise, dealing with the fragments in their severalty.'

This fragmentation has resulted, among other things, in a host of contradictory views among dietitians, each one of the different diets advocated possessing a company of followers ready to argue its exclusive merits with almost religious fervour. Thus you have the vegetarian; the fruitarian; those who never eat proteins and starch at the same meal; and those who stew all their foods together in the same pot; those that say you must drink before meals; others that believe in drinking after meals; those that drink between meals, and besides all these, and many more, there is the vast majority that eat what they want when they want it (or can get it) and drink when they are thirsty. This majority is given to labelling all the others as faddists, and indeed there does not seem to be a very noticeable difference in health between any of these groups. No wonder then, that the average person is apt to be a little sceptical when he is told that health depends on diet.

What then should be the reply to the would-be seeker after health who asks: 'What shall I eat that I may be whole?'

For an answer let us go to those people from whom Dr. Wrench, in his student days, felt instinctively that such knowledge should be sought, namely the ultra-healthy. Not the occasional individual of whom one says that he is 'abnormally healthy' (a revealing adverb) but to whole groups of people to whom a state of full health is normal.

Five such groups exist, or have existed, about which a good deal of statistical data is available. We will examine these in turn, noting how they live and what they eat, and see if we can discover among them a common factor of which it permissible to say—here lies the secret of health.

We will take first the people of Hunza, a small native state in the extreme northernmost part of India attached to the Gilgit Agency. The origin of these people is somewhat of a mystery. Both in physical characteristics and language they differ from their neighbours, and indeed from all the other peoples of the Indian subcontinent. Only one thing seems certain, that they have inhabited their valley since the extremely distant past. The massive stone walls, the building of which must have preceded their admirable terraced agriculture, have a parallel only in the masonry left by the Peruvian civilization which preceded the Inca conquest.

The Hunza valley is a gorge running east and west cleft in a towering mountain range. It is arid in summer and bitterly cold in winter, but owing to the system of irrigation, it is extremely fertile and an immense variety of fruits and vegetables are cultivated by these industrious people.

Something of their superb health and stamina has already been indicated in Chapter 2. All travellers passing through their valley speak of their outstanding physique, courage and good humour. McCarrison, who for some time was Medical Officer to the Gilgit Agency, has said of them:

'These people are unsurpassed by any Indian race in perfection of physique. They are long lived, vigorous in youth and age, capable of great endurance and enjoy a remarkable freedom from disease in general.

'During the period of my association with these people I never saw a case of asthenic dyspepsia, of gastric or duodenal ulcer, of appendicitis, of mucous colitis, of cancer. . . .

'Among these people the "abdomen over-sensitive" to nerve impressions, to fatigue, anxiety, or cold was unknown. The consciousness of the existence of this part of their anatomy was, as a rule, related solely to the feeling of hunger. Indeed, their buoyant abdominal health has, since my return to the West, provided a remarkable contrast with the dyspeptic and colonic lamentations of our highly civilized communities.'

They are admirable cultivators, 'far famed as such and "conspicuously ahead of all their neighbours in brain and sinew" stated Shomberg. Their big irrigation conduit, the Berber, is "famous everywhere in Central Asia. . . ."

'Amongst the peoples of the Agency not only are they "as tillers of the soil quite in a class apart, they alone—and this always strikes me as truly remarkable—are good craftsmen." As carpenters and masons, as gunsmiths, ironworkers, or even as goldsmiths; as engineers for roads, bridges or canals, the Hunza men are outstanding.'

The Hunza are favoured in their fertile valley, but their perfect health cannot be put down to the locality in which they live, for next door to them is another, and equally fertile, valley, also running east and west and separated from Hunza only by a 20,000-foot mountain wall. In this valley live the Ishkomanis. These people, 'though living under apparently like conditions to their neighbours, were poor, undersized, undernourished creatures. There was plenty of land and water, but the Ishkomanis were too indolent to cultivate it with thoroughness; and the possibility of bad harvests was not enough to overcome their sloth. . . . "They had no masons or carpenters or craftsmen in their country. Many of them showed signs of disease." '

We can thus rule out climate as the secret of the Hunza health. Now, let us look at their mode of life and their diet. For ten months of the year they can be said to live in the open air, for men, women, and children work in the fields. They remain mainly indoors during the period of severe winter storms, but their houses are better, and better ventilated than those of their neighbours, their sanitation is also better and follows the 'immemorial custom of the Far East'. Unusual care is also taken to protect their drinking water by storing it in separate covered cisterns.

Their diet is a very varied one. It consists of 'wheat, barley, buckwheat, and small grains; leafy green vegetables; potatoes (introduced half a century ago), other root vegetables; peas and beans, gram or chick pea, and other pulses; fresh milk and buttermilk or lassi; clarified butter and cheese; fruit, chiefly apricots and mulberries, fresh and sun-dried; meat on rare occasions; and sometimes wine made from grapes. Their children are breast fed up to three years, it being considered unjust to the living child for its lactation to be interrupted by a maternal pregnancy.

'The Hunza do not take tea, rice, sugar or eggs. Chickens in a confined area destroy crops and are not kept.'

Except for the wider range of small grains and the very occasional meat, this closely resembles any European lacto-vegetarian diet, and, at first sight, seems to support the view so widely held by nutrition experts, McCarrison among them, that of all diets the lacto-vegetarian is the healthiest. That it is a good diet is incontestable; it will, however, presently be shown that there are other peoples whose health and stamina is equal to that of the Hunza whose diet is the very opposite of theirs.

The Hunza foods then are not unlike our own, but there are important differences in the normal methods of preparation and cooking. Both we and the Hunza are great bread-eaters, and both prefer wheat bread, but the Hunza wheat is eaten freshly ground, and the unleavened bread made with it invariably contains the whole of the grain, with its vital germ and its protective skin, both of which are removed in the process of milling white flour.

Dr. Wrench, writing on the properties of skin in general, points out that skin does not protect only in a mechanical way as a mere covering, but in a living way. All skins 'can regrow themselves if injured, and beneath and within them they store substances upon which they can call to strengthen their efforts'. The value of bran, the skin of wheat, is well known to all stock feeders. All carnivorous animals relish the skins of their prey. The Greenlanders, Wrench points out, eat the skin of the narwhal: 'The Chinese and other peoples also eat the skins of animals and birds. Everything living has a skin of some sort to protect it. It protects it by its extra toughness, but also if microbes and other minute enemies do attack, it is there on the frontier that the battle is waged. In and near the skin are marshalled the protective forces. Any creature that eats the skin of vegetable, fruit, or animal, also eats these protective materials marshalled on the frontier, and may benefit in its own protection thereby. Whether such a pretty hypothesis is true or not, there are suggestions that skins possess a peculiar value. . . . The skin and adjacent part of the potato is the best part, as the Irish know. So also is it the case with the carrot, and, it is said, with young marrows, cucumbers, gherkins, artichokes, radishes, and celery. There is, therefore, a little evidence for the hypothesis.'

A fondness for skin is an outstanding feature of the Hunza, they do not peel their vegetables, or wash and soak them to the extent we do. Vegetables play a great part in their diet and are very commonly eaten raw. 'They are fond of raw green corn, young leaves, carrots, turnips, and, as it were to exaggerate their veneration for freshness, they sprout their pulses and eat them and their first green. This eating of sprouting pulse or gram is widespread in northern India, and undoubtedly within it there is a health which there is not in the pulse itself.'

Fuel in Hunza is scarce, and when they do cook their vegetables they are boiled in covered pots as is the usual habit in this country.

'But the process is more comparable to our way of steaming and cooking in their own juice. Very little water is added. When this has been used up more is added. The water in which the vegetables are cooked is drunk either with the vegetables or later. The point is that it is part of the food. It is not thrown away.'

This taking of vegetable water is obviously sensible, for many of the valuable mineral salts which vegetables contain pass into the water in which they are cooked, particularly if the vegetables are peeled before cooking. 'There is abundant evidence from the scientists of the loss that occurs through the throwing away of vegetable water of phosphorus, calcium, iron, iodine, sulphur, etc. Quite a considerable proportion of the pharmacopœia seems to have arisen owing to this waste. Quite a considerable number of the doctors' prescriptions and patent medicines may be due to the need to replace the salts of the food in those who suffer from this loss. The similarity of the medicines and the lost salts is too close for one not to be profoundly suspicious that the methods of cooking cause or contribute to the subsequent need of the medicines.'

The Hunza drink milk in considerable quantity, they drink it whole and they boil the fat from it to form clarified butter or ghee, which they spread on their bread and also use for cooking. They drink the buttermilk which remains, and both this and their whole milk they preserve in hot weather by souring. Meat is a 'rare pleasure', most of their livestock being dairy animals. They rarely eat meat more than once in ten days, and often only about once a month. When they do, they eat all that is edible in the carcass and stew it together with their vegetables and pounded wheat. But ranking above all the foregoing in the Hunza diet is fruit. ' "The Hunza are great fruit eaters, especially of apricots and mulberries. They use apricots and mulberries in both the fresh and dry state, drying sufficient of their rich harvest of them for use throughout the autumn and winter months." (McCarrison.) They eat the fruit fresh in season, cracking the stones and eating the kernels as well. Otherwise they take them, particularly sun-dried apricots, and eat them as they are, or rub them in water to form a thick liquid called *chamus*. Dried mulberries they put into cakes as we do sultanas. They do not cook their fruits. "Fruit is really the Hunza staple. It is eaten with bread, far more so than vegetables, as it is more abundant." ' (Schomberg.) . . . ' "Even the animals," said Durand, "take the fruit diet, and you see donkeys, cows and goats eating the fallen mulberries. The very dogs feed on them, and our fox-terriers took to the fruit regimen most kindly and became quite connoisseurs." ' They ferment some of their fruit juices and on festive occasions drink their own home-made wine.

So far the main differences between the Hunza diet and our own seems to be that the Hunza foods are all natural foods, they are eaten fresher than ours, and they are consumed whole, but there is one more difference, the most fundamental of all, and this lies in the way in which these foods are grown.

In their system of agriculture which has been continued 'century after century' the chief factors in their plant food have been two.

'Firstly, there is the continuous slight renewal of the soil by a sprinkling of the black glacier-ground sand, which is brought to the fields by the aqueducts.

'Secondly, there is the direct preparation by man of food for the plants, given in the form of manure.

'The Hunza, in their manuring, use everything that they can return to the soil. They carefully collect the cattle manure and store it in the byres. They collect all vegetable parts and pieces that will not serve as food to either man or beast, including such fallen leaves as the cattle will not eat, and mix them with the dung and urine in the byres. They use the human sewage after keeping it for six months. They take silt from special recesses built in their irrigating channels. They collect the ashes of their fires. All these they mix together and make into a compost. They also spread alkaline earth from the hills on their vegetable fields on days when the fields are watered.'

O. F. Cook of the Bureau of Plant Industry of the US Department of Agriculture has written: 'Agriculture is not a lost art, but it must be reckoned as one of those which reached a remarkable development in the remote past and afterwards declined.' As an example he cites the system of the ancient Peruvians which enabled them to support large populations in places 'where modern farmers would be helpless'.[1]

Travellers who have visited both Peru and the North-West Indian Provinces have been struck by the resemblance between the stone aqueducts and mammoth walls that support the terraced fields of both areas. Describing those of Hunza the late Lord Conway wrote: 'The path that leads up to Baltit[2] is bordered on either side by a wall of dry cyclopean masonry the undressed component parts of which are very large and excellently fitted together . . . a monumental piece of simple engineering . . . the valley between the cliffs and the edge of the river's gorge is covered with terraced fields. . . . The cultivated area of the oasis is some five square miles in extent. When it is remembered that the individual fields average as many as twenty to the acre, it will be seen what a stupendous mass of work was involved in the building of these walls and the collection of earth to fill them. The walls have every appearance of great antiquity, and alone suffice to prove the long existence in this remote valley of an organized and industrious community. . . .

'To build these fields was the smaller part of the difficulties that husbandmen had to face in Hunza. The fields also had to be irrigated. For this purpose there was but one perennial supply of water—the torrent from the Ultar glacier. The spout of that glacier, as has been stated, lies deep in a rock-bound gorge, whose sides are for a space perpendicular

[1] 'Staircase Farms of the Ancients', *National Geographic Magazine*, May 1916.
[2] The capital of Hunza.

cliffs. The torrent had to be tapped, and a canal of sufficient volume to irrigate so large an area had to be carried across the face of one of these precipices. The Alps contain no *Wasserleitung* which for volume and boldness of position can be compared to the Hunza canal. It is a wonderful work for such toolless people as the Hunzakats to have accomplished, and it must have been done many centuries ago and maintained ever since, for it is the life-blood of the valley.'

Thus it can be seen that the Hunza appear to form a direct link between the present day and that 'remote past' in which agriculture reached such a 'remarkable development'. They are a people perhaps as ancient as the Incas, but who, unlike the Incas, have survived, and in their survival have preserved their ancient lore, and in the preservation of that lore have preserved the wholeness of their health and that of their crops and livestock, which Dr. Wrench tells us is on a par with their own.

The blight which Western civilization usually casts on such people, has so far escaped them. Whether it will continue to do so is another matter. Since they have come under British suzerainty their population has increased from about six thousand to fourteen thousand, and this has resulted in a shortage of food in the pre-harvest period:

'Colonel D. L. Lorimer, who was Political Agent at Gilgit, 1920–4, and revisited the Hunza and lived amongst them at Alibad, 1933–4, four miles from the capital, Baltit, told me that not only did they seem smaller to him at his second visit, but that the children appeared under-nourished for the weeks preceding the first summer harvests half-way through June; and, moreover, that the children suffered at that time of the year from impetigo, or sores of the skin, all of which vanished when the more abundant food came.'

A sign, incidentally, that it is not by virtue of their race, or habitat, or housing, that they are normally immune from bodily ailments.

We will now go from Latitude 37 to the northern and Arctic regions: from the lacto-vegetarian diet of the Hunza to the carcass diet of the islanders of Faroe, Iceland, and Greenland. Early records of these peoples show them to have been every whit as healthy as the Hunza, yet these people are, or were, almost entirely carnivorous.

'These Danish possessions are three isolated lands from which no Western civilized person would expect to glean wisdom. But, as we have already seen in the case of food and health, isolation locks up the most valuable secrets. The peoples of these three lands, living either near or actually within the Arctic Circle, offer in three degrees, from Faroe to Greenland, an increasingly animal-fish-bird diet. It must not be called a meat diet; that is inaccurate as will be seen. It was largely a diet from the sea, and with the great health of the sea, a "soil" outside the realms of terrestrial man.

'The diet of the Faroe Islanders, when they were more isolated than now, was given in a book published by the Edinburgh Cabinet Library in 1840. It was mainly a whole-carcass diet of animal, bird and fish. The islanders ate not merely meat, but everything that could be eaten. There was no such thing as offal. They also made the carcasses gamey by hanging for weeks and even months. In addition to their whole-carcass food they had barley meal, unleavened barley bread, a few vegetables such as cabbages, parsnips and carrots.

'They drank milk, beer, and, on festive occasions, brandy. But the main food was animal, bird and fish.

'The islanders numbered a few thousands, were of the same origin as the Icelanders, and were, "in general, remarkably intelligent. They are extremely healthy, and live to a great age, and an old man of ninety-three years lately rowed the governor's boat nearly ten miles." One danger they incurred was an epidemic catarrhal fever, such as we call influenza, which "prevails after the arrival of the ships from Denmark in the spring", after the winter's scarcity. It spreads rapidly and was sometimes fatal. Otherwise, "but few diseases are prevalent amongst them".

'The inhabitants of Iceland offer a similar and even more interesting picture of carcass diet. McCollum and Simmonds, in *The Newer Knowledge of Nutrition* (1929) summarize the chief facts. "This island was settled in the ninth century by colonists from Ireland and Scandinavia, who took with them cattle, sheep and horses. Their diet was practically carnivorous in nature for several hundred years. Martin Behaim (quoted by Burton), writing of Iceland about 1500, stated: 'In Iceland are found men of eighty years who have never tasted bread. In this country no corn is grown, and in lieu, fish is eaten.' "

'Burton, quoting Pearse, states that rickets and caries of the teeth were almost unknown in Iceland in earlier times. . . . The health conditions were good and dental caries was unknown until after 1850. Stefansson exhumed ninety-six skulls from a cemetery dating from the ninth to the thirteenth centuries and presented them to Harvard University. They have been described by Hooton (1918), who found no evidence of caries in any of them. There were but three to four defective teeth in the entire series, and these had suffered mechanical injury. During the last half century caries has steadily increased in Iceland.

'Modern Iceland had not the isolation of the period which Burton described. There had been a great advance in civilization and population. Fifty per cent of the people now live in towns or trading stations. There are four agricultural schools. Potatoes, turnips, and rhubarb are cultivated. Iceland imports the trade foods, such as flour, sugar, preserved fruits and tinned foods. Caries has become common, as have many other ailments.'

That this regrettable decline in health cannot be attributed to the change from country to town life, is proved by a remarkable experiment carried out in Denmark itself during World War I.

'The blockade, following the entry of the USA into the war, put the Danes in a very serious position. Professor Mikkel Hindhede, Superintendent of the State Institute of Food Research, was made Food Adviser to the Danish Government to deal with it.

'The problem that faced him was this; Denmark had a population of 3,500,000 human beings and 5,000,000 domestic animals. She was accustomed to import grains from the United States for both. There was now a shortage of grain foods.

'In this crisis Hindhede decided that a drastic reduction in the livestock must be made. So some four-fifths of the pigs were killed and about one-sixth of the cattle. Their grain food was given to the Danes, and it was given . . . as wholemeal bread with the extra coarse bran that is not put into ordinary wholemeal bread, incorporated.

'In addition to this bread, or Kleiebrot, which was made official for the whole country, the Danes ate porridge, green vegetables, potatoes and other root vegetables, milk, butter, and fruit. No grain or potatoes were allowed for the distillation of spirits, so there were no spirits. Half the previous quantity of beer was permitted.

'As some pigs were left, the people on the farms got meat; the people in the cities—40 per cent of the population—got very little meat. Only the rich could afford beef.

'The food regulations were begun in March 1917 and were made stringent from October 1917 to October 1918.

'The result of this enforced national diet was a remarkable lowering of the deathrate. The deathrate, which had been 12·5 in 1914, now fell to 10·4 per thousand, "which is the lowest mortality figure that has been registered in any European country at any time". (Hindhede.)

'Hindhede puts this impressive result in another way. Taking the average from 1900 to 1916 as 160, in the October to October year it was 66. Even in men over sixty-five the figure fell to 76.

'Hindhede attributes this extraordinarily rapid and marked change to two things: (1) less meat, (2) less alcohol. He regards the bran as having largely filled the gap of the scanty or absent meat, bran having a good proportion of vegetable meat or protein. He regards the experiment as a triumph for his previous teaching. "The reader knows," he writes in the *Deutsche Medizinische Wochenschrift* of March 1920, "how sharply I have emphasized the advantages of a lacto-vegetarian diet. I am not in principle a vegetarian, but I believe I have shown that a diet containing a large amount of meat and eggs is dangerous to health." '

And yet we have only to turn to another Danish possession (ironically enough) to find a refutation of this rather narrow view.

'The north-west coast of Greenland, where the Polar Eskimos live, is within the Arctic Circle. It is the most isolated and the least affected by civilization of these three possessions of Denmark.

'Some attempts at gardening have been introduced by the Danes, but previously the only vegetable food the Eskimos got was from the profuse but, in species, limited vegetation of the Arctic summer. Otherwise they lived mainly on sea animals and sea birds. There was no offal. They ate everything that could be eaten. When it was frozen they often ate it raw. The thick, heavy skin of the narwhal is particularly favoured. The millions of sea birds which visit their coast supply a winter store of meat and eggs.

'The Eskimos are also exceptionally healthy. "The fact that the Eskimos of this polar tribe have such excellent physique, hair, and teeth, and such superb health without any trace of scurvy, rickets, or other evidence of malnutrition," write McCollum and Simmonds, "is interesting in the light of their restricted and simple diet."

'It is also interesting as a counterweight to Hindhede and other nutritionists who plump for the excellent lacto-vegetarian diet. There are other excellent diets, and the whole-carcass one of the Polar Eskimos is one of them.'

From the far north our next jump is to the Island of Tristan da Cunha in the South Atlantic. 'The people of this island are people of our own race living on the products of sea and soil, most of them have perfect teeth which last them all their lives. . . .

'Mr. James R. A. Moore, LDS, RCS (Eng), visited the island in 1932 and again in 1937. In 1932 he examined 156 persons and 183 in 1937. Of the 3,181 permanent teeth in the former year, there were 74 carious and of the 3,906 in the latter year there were 179 carious.

'He speaks of the physique of the people as being good. They are well set up, clean and well nourished. The children are breast fed and are not weaned until at least one year old. Fish and potatoes are the staple diet, meat occasionally, milk and butter sufficient. Eggs form a big item of the island diet and are mainly Mollyhawk and penguin. Vegetables are not plentiful, but beetroot, lettuce, beans and onions are now being grown. Imported flour and sugar are regarded as luxuries, but they have been brought in to a greater extent latterly, which may account for the tendency of the teeth to deteriorate.

'The fat in adequate amount is provided by rendering down the carcasses of young Mollyhawks and petrels and is used extensively for frying. Sea water is evaporated to provide salt.'[1]

It will be seen that the people of this island, also noted for their sound health,[2] have a more varied diet than that of the Eskimo, though like them much of it is derived from the sea. Eggs form a large part of their diet—one of the items condemned by Hindhede. It is worth noting that a marked difference exists between the methods of cooking adopted by the Tristan Islanders and the Hunza, for whereas the latter cook everything together in one pot, the people of Tristan never partake of more than one kind of food at the same time.

For our fourth example of a (once) superlatively healthy race, we must go to the North American Indian of the pioneer days. Observation of these people since they have been forced to live in reservations has been very carefully recorded.

'All who observed the Indians in their primitive state agree that most of them were exceptional specimens of physical development. With few exceptions, however, during two generations, they have deteriorated physically. The reason for this is apparently brought to light by a consideration of the kind of food to which they have restricted themselves since they have lived on reservations.

'There is no group of people with a higher incidence of tuberculosis than the non-citizen Indian. As wards of the Government they have been provided with money and land, but have in general shown little interest in agriculture. They have lived in idleness, and have derived their food supplies from the agency stores. In addition to muscle cuts of meat they have, therefore, taken large amounts of milled cereal products, syrup, molasses, sugar and canned foods, such as peas, corn, and tomatoes. In other words, they have come to subsist

[1] Ref. to 'Medical Testament'.

[2] For further information on health conditions in Tristan da Cunha see Erling Christophersen, *Tristan da Cunha*, English translation published by Cassell, London, 1940.

essentially upon a milled cereal, sugar, tuber, and meat diet. On such a regimen their teeth have rapidly become inferior and badly decayed. They suffer much from rheumatism and other troubles which result from local infections. Faulty dietary habits are in great measure to be incriminated for their susceptibility to tuberculosis.

'Other classes of Indians, who have become successful farmers, have not deteriorated as a result of contact with civilization, except in so far as they have suffered from alcohol and venereal infections. The non-citizen Indian has suffered, not because of contact with civilization, but because he has been forced into dietary habits which are faulty.' (McCollum and Simmonds.)

In the days of their prime these people subsisted mainly on the wild game of virgin forest and prairie, regions in which, as was pointed out in a previous chapter, the law of return operated fully.

For the last example I go back to the continent of Asia, to the people of rural China. Their diet is nearer to that of the Hunza than to any of the other examples we have looked at. Fruit, vegetables and sprouted grain are staples of both diets, but unlike the Hunza the principal cereal of the Chinese is rice, not wheat, and they also eat meat, birds, fish and eggs. They are in addition, as is well known, great tea drinkers. In common with the other four groups they eat the whole carcass and the whole grain or vegetable. In the matter of preparation they resemble the Hunza in that everything is eaten together. But are we justified in claiming that they are healthy?

Sooner or later all advocates of organic farming cite the Chinese, going so far as to call them the fathers of good husbandry. Their authority for doing so is almost always Professor King's famous book, *Farmers of Forty Centuries*. The critics of the 'organic school', however, challenge the accuracy of King's report. They say that he was only in the country for a few weeks, that the Chinese people as a whole have an abnormally high deathrate, and that the whole country is riddled with disease, most of which is sewage borne because the peasants *fail* to compost their human wastes.

I have made great efforts to check these two opposing views. The truth seems to lie in the statement from Lord Northbourne's *Look to the Land*, already quoted in the first chapter of this book, that 'China presents remarkable contrasts between the best and the worst'. It is a vast country. Undoubtedly conditions are very bad indeed in some areas, particularly in the overcrowded cities, both as regards health and sewage disposal. But there seems equally little doubt that in certain rural areas composting of a very high order, amounting to a fine art, has been practised for centuries. But, however the overall picture should be painted, one fact seems indisputable, namely, that through the operation of the closed cycle (i.e. without the importation of chemical fertilizers) the soil of China has—despite periodic floods and famines—supported a huge population and a high culture for a period of 4,000 years. For this reason, I feel quite justified, after having drawn attention to the other side of the picture, in once more quoting King. He was, after all, Chief of the Division of Soil Management in the United States Department of Agriculture, and as such a qualified observer, recording facts as he found them. The references to the 'Medical Testament' contain an admirable summary of King's findings, and it is from that document that the following account is taken.

'King frequently inserts into his pages the cheerful, vigorous and healthy appearance of the Chinese lower classes, the Shanghai coolies, "fully the equal of large Americans in frame, but without surplus flesh"; "their great endurance", "both sexes are agile, wiry, and strong" (Hong Kong); "lithe, sinewy forms, bright eyes and cheerful faces, particularly among the women, young and old" (Canton); "everywhere we went in China the labouring people appeared healthy and contented, and showed clearly that they were well nourished". Cheerfulness is, indeed, common to those peasantries who follow the old agricultural ways.'

'The average of seven Chinese holdings . . . indicates a maintenance capacity of 1,783 people, 212 cattle or donkeys and 399 swine—1,995 consumers and 399 rough food transformers per square mile of farmland. These statements for China represent strictly rural populations. The rural population of the USA in 1900 was placed at 61 per square mile of improved farm land and there were 30 horses and mules. . . .'

'They [the Chinese] have long realized that much time is required to transform organic matter into forms available for plant food, and, although they are the heaviest users in the world, the largest portion of this organic matter is predigested with soil or subsoil before it is applied to the fields. This is at an enormous cost of human time and labour, but it practically lengthens their growing season and enables them to adopt a system of multiple cropping which would not otherwise be possible. By planting in hills and rows with intertillage it is very common to see three crops growing upon the same field at one time, but in different stages of maturity—one nearly ready to harvest, one just coming up, and the third at the stage when it is drawing most heavily on the soil. . . .'

This disposes of the theory that increased production and heavy cropping have been responsible in this country for our diseases in crops. The Chinese have been cropping in this way for forty centuries.

'The Chinese manure or compost is made of everything that can be collected which once got its life from the soil, directly or indirectly. They are mixed together until they form a black friable substance which is readily spread upon the fields. King describes a number of different processes he saw in different parts of China. One he describes as being carried out in compost pits at the edge of a canal, a process entailing "tremendous labour of body and amount of forethought". For months before his visit men had brought waste from the stables of Shanghai, a distance of fifteen miles by water. This they had deposited upon the canal bank between layers of thin mud dipped from the canal, corresponding to silt collected in and taken from the recesses in the Hunza aqueducts, and left to ferment. The eight men at King's visit had nearly filled the compost pit with this stable refuse and canal silt. The pit was in a field in which clover, with its peculiar power of taking nitrogen from the air, was in blossom. This was to be cut and piled to a height of five to eight feet upon the compost in the pit, and also saturated layer by layer with canal mud. It would then be allowed to ferment twenty to thirty days, until the juices set free had been absorbed by the winter compost beneath and until the time that the adjacent land had been made ready for the coming crop. The compost would then be distributed by the men over the field.

'At another time he saw a compost pit within a village in which had been placed all the manure and waste of the households and streets, all stubble and waste roughage of the fields, all ashes not to be applied directly, mixed up with some soil. Sufficient water was added to keep the contents of the pit saturated and to promote their fermentation. All fibres of organic material have to be broken down, which may require working and reworking, with frequent additions of water and stirring for aeration. Finally the mixture becomes a rich complete fertilizer. It is then allowed to dry and is finely pulverized before it is spread upon the land.

'Every foot of land, says King, is made to provide food, fuel, or fabric. "The wastes of the body, of fuel, and fabric, are taken back to the field; before doing so they are housed against waste from weather, intelligently compounded and patiently worked at through one, three or even six months, in order to bring them into the most efficient form to serve as manure for the soil or as feed for the crop." '

These then are the five peoples[1] which either still enjoy an exceptional measure of health, or else until very recently have done so. What have they in common?

Not race, for the groups include white, brown, red, and yellow races.

Not climate—there could hardly be greater contrasts than between the plains and hills of rural China or the prairies of North America and the precipitous mountain crags of the northern provinces of India, or than between the frozen north and the luxuriant warmth of Tristan da Cunha.

Not diet—in the ordinary sense—for these range from the lacto-vegetarian diet of the Hunza to the almost purely carnivorous diet of the Eskimo, with almost every variant in between.

Not methods of preparing their food either, for though there are certain resemblances—as between Hunza and Chinese for example—no methods are common to all five.

In fact it seems clear that it is not in *kind* at all that we must look for our common factor, but in *quality*.

All five groups have good air to breathe, but that cannot by itself be the secret of their health, or our own hill and country dwellers would have health to compare with theirs, which, unfortunately, they have not.

The only discernible common factor, other than good air, seems to be that the diets of all five groups are 'whole' diets in the full sense of the word. That is to say: (a) every edible part contained in the diet is consumed; (b) in every case the foods are grown by a system of returning all the wastes of the entire community to the soil in which they are produced. For the sea, too, is a 'soil' in this sense, supporting its teeming population by means of the rule of return—the everlasting cycle of life and decay; (c) all the foods are natural unprocessed foods; (d) the diets start before life begins; the parent is as healthy as the child.

There is a complete and continuous transference of health from a fertile soil, through plant and/or animal to man, and back to the soil again. The whole carcass, the whole grain, the whole fruit or vegetable, these things fresh from their source, and that source a fertile soil.

[1] It appears that to this list should be added the people of Prince Edward Island. See report by Dr. Enid Charles in the Canadian *Journal of Economics and Political Science*, VIII, 1942.

Herein appears to lie the secret. If this be true, then the answer to our question, put at the beginning of this chapter, would appear to be that *any* diet is a health-promoting diet so long as it conforms to these three rules, and the first of these is a fertile soil.

'The importance of the method of culture of food is primary, radical, and fundamental in the matter of health. It exceeds all other aspects of nutrition—if, that is, one separates any aspect of what is a whole.'

In the case of diets based on agriculture, such a view brings us back again to humus farming.

[It also served to emphasise the need for further investigation, and so led to the establishment of the Haughley Experiment.]

CONNECTING LINK

Being an Epilogue to Part I and a Prologue to Part II

Connecting Link

It was in the mid 1930s that I began to study the scattered evidence which I later put together in book form.

From the outset, the case, as it built up, seemed to me, and to many others with whom I discussed it, to be crying out for further investigation by means of long-term, field-scale, fundamental ecological research. This would involve a controlled experimental programme totally different from any hitherto set up by established agricultural research stations.

The need for such a project was fully endorsed by top-ranking men in both medicine and agriculture, but in spite of this it soon became apparent, from the attitude of the establishment of the day, that prospects of any official body being persuaded to launch such an undertaking, in the foreseeable future, were remote. Those of us pressing for this were thus driven to the conclusion that pioneering, even in matters of health, must be the function of individuals.

We were, I think, somewhat influenced over this by the PEP report on the British Health Services of that period (1936). This declared: 'Unquestionably, many of the greatest achievements of the British Health Services up to the present time have been due to voluntary initiative.' And it continued: 'The scope for work by voluntary bodies in the health services appears to fall into five classes: (a) work in which freedom of opinion in teaching or research is absolutely vital; (b) work which is of a controversial nature; (c) work which is of a pioneering character and contains an element of risk; (d) other "charitable" work which no public body thinks part of its duty although its value is generally recognized; (e) work where the disease or health problems involved affect the social life of the individual and there is scope for personal influence or example.'

Luckily great causes usually produce their pioneers. First must come those who make the initial discoveries, and after them men of vision, prepared, in the face of all difficulties, to prove that the discoveries are worth official recognition. The 'cause' of exploring the relationship between humus and health, with its momentous implications, was no exception—the indomitable pioneer in this case was the late Miss Alice Debenham, though, tragically, she died at the very outset of her great purpose, leaving to others the task of bringing to fruition the seed she sowed.

A practical farmer, trained in science and medicine, and during the latter years of her life an invalid, Alice Debenham saw very clearly the potential importance of the evidence concerning soil fertility and health. She saw equally clearly that this scattered evidence must be collected and reproduced under controlled conditions if it were to convince the scientific world, and that unless science is convinced, governments will not act.

Outstandingly public spirited, she founded a Research Trust to carry out this work. As custodians of it, she appointed the East Suffolk County Council. By this means the Trust immediately obtained an official status that a purely private body would not have achieved, while, at the same time, the terms of the Trust gave it a much greater freedom of action than would have been possible to a government-run concern. To this Trust she presented, under a deed of gift, about 80 acres of farmland, at Haughley in Suffolk, together with a limited number of farm-buildings and an admirably modernized farm-house. These 80 acres marched with a 156-acre holding which I had been farming since 1919 and it was agreed between us that this land should be leased or sold to the Trust so that the total working acreage of the experimental area should be close on 200 acres. This was considered to be the minimum acreage needed for carrying out the experiment for which the Trust was formed. The two properties, New Bells Farm (156 acres) and Walnut Tree Farm (80 acres), together formed a compact unit. Except for 20 acres, let off, I had been farming the two as one farm since the early 1930s (and New Bells Farm since October 1919), which provided a fairly long period of known history. The characteristics of the area are: (a) flat land subdivided by hedges and deep ditches into small fields (average 6–8 acres), and (b) a soil which is a nearly uniform alkaline clay-loam, overlying glacial formation of clay and flint with sand pockets, which in its turn overlies chalk. The uniformity of these characteristics throughout both properties seemed to make them ideally suited to the project.

Under the original Trust deed, the East Suffolk County Council's acceptance of Custodian Trusteeship was an act of vision the credit for which belonged to its Clerk, the late Sir Cecil Oakes, CBE, who remained an invaluable guide and friend of the Experiment from its inception to the day of his death in 1959.

The Trust was formed in 1939, and a subsidiary trading company, Haughley Research Farms Ltd., was registered to operate the farms. Some of the consultants were also appointed. The proposed plans of operation, and the evidence which gave rise to them, were then printed in pamphlet form. In a preface to this pamphlet Sir Robert Hutchison[1] wrote: 'It is highly important that the question should be settled without delay and the experiments planned to be carried out at Haughley are admirably calculated to determine the question at issue once and for all. For that reason they are deserving of every encouragement and support from the medical profession.' Sir George Stapledon wrote in much the same vein for agriculture. This pamphlet was privately circulated, with a view to obtaining other expert opinion, and the result was so encouraging that a campaign to raise funds was about to be launched when the war started.

During the war it was impossible to raise money for private research, and war-time restrictions on farm production also made it impossible to initiate the livestock programme, because crops had to be grown for human consumption. A most welcome measure of recognition that the farms were undertaking research was, however, given in that they were *exempted from orders as to fertilizer application.* Thus the basic soil treatment plan for the three sections (described in Part II) was adhered to throughout the war. This provided the necessary initial pre-conditioning of the fields in readiness for the first true preparatory period, which began in 1948.

In the meantime some publicity for the work had been obtained by the publication by Faber & Faber in 1944 of my book *The Living Soil* which, in addition to Part I of the present volume, contained a chapter describing the Haughley Research Project, with a report on the progress of its preparatory work to date. The response to this book indicated that the research plan had a considerable amount of support from many widely scattered individuals. Nevertheless, insufficient funds were forthcoming in the immediate post-war period to enable the Trustees to take the decision to continue the Experiment, so that it appeared that it would be still-born. This was in 1947. The year before that, however, a group of people, representing a wide cross-section of the population, had met together to form the Soil

[1] Sir Robert Hutchison, Bart., MD, LLD, DSc, FRCP, President of the Royal College of Physicians.

Association, a non-profit-making society, which now has a world membership. Its objects are summarized under three headings:

(a) To bring together all those working for a fuller understanding of the vital relationships between soil, plant, animal and man.

(b) To initiate, co-ordinate and assist research in this field.

(c) To collect and distribute the knowledge gained so as to create a body of informed public opinion.

When the Haughley Trustees decided that they could not carry on, they offered the assets of the Trust to this newly formed organization. Although only in its infancy, the Soil Association felt that this was a challenge that they could not refuse. Supported only by faith, they accepted, and appointed a committee of consultant scientists to plan the next stage. This is the point at which Part II of the present book takes up the story.

To complete this connecting link between Parts I and II, which represents not only a jump in time, but also one in ecological thinking, I asked Dr. Innes Pearse (co-author of *The Peckham Experiment*, widow of one of our chief consultants—the late Dr. G. Scott-Williamson—and herself a research biologist) to write something with special reference to the Haughley Experiment, to explain to my readers what is meant by *fundamental* (or *basic*) research, and why, in the field of biology (i.e. life) the methodology of such research (concerned with quality of *living*) must differ so completely from that employed in *applied* research (concerned with quantitative measurements). This she most kindly consented to do.

It is not easy reading, but then it is not an easy subject. If it were, it would not have been necessary to seek her contribution towards clearing up the confusion concerning it which arose as the experiment progressed and developed.

It is my hope that if any readers of Part II lose sight of the significance of this difference (which is immensely important to an understanding of the Haughley story) they may find it helpful to refer to what I consider the masterly exposition which follows.

E.B.B.

Basic Research

The Haughley Experiment was planned to inquire into the importance of the vital attributes of the soil for the health of plant, animal and of man. It was seen from the outset that the subject could only be understood in the sense of a *whole*. This approach, both to the nature of the soil and to health, was a revolutionary one at the time when the need for such an inquiry was first presented in *The Living Soil*.

The procedure essentially involved what today would be called ecological research. That means study of the functional relationships within an assemblage of organic entities: man, animal, plant along with all the teeming hierarchical society of living inhabitants of the soil.

It was conceived that the balance of exchange in the ecosphere that encompasses them all in the continuity of living, is sustained in *mutuality of action*; each taking what it needs and rejecting what it has had no use for, thereby sustaining the needs of others within that inhabitation. As a shift occurs through the action of one, so all shift within the functional organization of the *whole*.

But more than this. What each utilizes in building up its own substance and carrying out its proper function, it stamps with the pattern of its own specificity; its own 'individuality', or uniqueness.

In the traffic of exchange there are then to be sought different types of contribution within the whole. There is that which is of specific pattern; and that, too, which is 'anonymous' and in use common to all. 'Heat', for example, generated in any transaction passes 'unlabelled' in its going, while there is that which having passed through the living organism, when ejected into the traffic stream, is imprinted with its specific identity; and leaving there its imprint on the scene for us to find—if we care to look!

This was the approach which gave shape to the Haughley Experiment. It was clear from the beginning that nothing less could suffice than a study of the functional relationships of the contained entities within the functioning 'whole'; and study of the contribution that its myriad inhabitants yield to the nutrition of all that grow within the functional ambit of that whole.

These living entities, animal, plant, insect together with the unseen multitudes that inhabit the soil, while drawing their every need from that environment, bring about weighty transformation in the soil. In these latter days we know something of their prodigious numbers; how, too, some of these transformations are brought about with their co-workers—maybe, for example, by one special enzyme for each chemical constituent used to fulfil a special need! This unceasing traffic mounts up, then, not only to a quantitative content—which is measurable; but also to a qualitative diversification, each item bearing its pattern of specificity. Though we may recognize these, they are still beyond our methodology for measurement.

The inhabitation of plant, animal and man, the living medium to be investigated, is outstandingly a qualitative one; it furnishes the ecosphere in which all live, with a liberal specific diversity of content; as well also as with a measurable quantitative content. Neither of these are to be ignored. Both have their implication not without significance for agriculture. So for instance where the inhabitation in which plant or stock is reared, is treated with chemical adjuvants, 'protective' pesticides, or selective poisons in relatively massed quantities, the diversity of the qualitative content and the mutual relationships that sustain the ecological balance, must thereby be seriously interfered with. It is not then enough to draw conclusions from quantitative estimation and analysis of the immediate products taken from the soil. The resultant findings from such analysis have to be seen in conjunction with changes that have coincidentally arisen in the soil and in its inhabitants; and seen also in comparison with other similar environmental situations sustained more naturally in continuity; and season by season.

If then for example, by substitute methods in the course of research or agricultural practice, we see evidence of increase in size or of the quantity of some product, it is not to be assumed that such product is 'better', more desirable, more worth growing in the belief that its consumption will sustain the health of plant, of beast or of man himself. Indeed, in the case of human health, we are well aware that increase *in weight* and/or in rapidity of growth and maturing, are by no means an infallible indication of increase in health! The finding may be no more than the first signs of disorder from an excess of feeding (as, for example, with sugars and fats), the increase in fact representing a differential *deficiency* of essential nutritional elements absent from the diet where a (natural) diversity should exist.

In a cursory fashion the above considerations may serve to illustrate the difference between basic research and applied research in this field. The analytical approach is capable of quick return: the approach through the whole, necessitates a long-term investigation. We may know something of the conditions that sustain the health of man: but given a quantitative increase in yield of produce from the soil, we are careless of the circumstances that sustain *its* 'life' in its rich diversity.

The path on which the Haughley Experiment set out was essentially long-term research, which could provide a backcloth to give balance, orientation and ultimate significance to controlled, but more partial, excursions into all forms of applied research in the field of agriculture, nutrition and health. Nothing less than a long-term inquiry could suffice for its purpose; for the abundance, resiliance and versatility of nature is so great that time—and the seasons, are required even to demonstrate the first indications of exhaustion expressed as recognizable deficiencies of the soil or of the nutrients it provides.

The Haughley Experiment has come to an end. But two full cycles of agricultural rotation conducted with a unique consistency in the treatment of the land have sufficed to throw up some results which were expected; and others quite unexpected. These leave no doubt as to the need for exploration of the subject *as a whole.*

The twenty years or more during which consistent conditions were maintained have shown the possibility of preserving the fertility of a closed cycle in a functioning whole. Meanwhile, since the experiment began, the rationality of the basis on which it was grounded has been considerably substantiated by the progress of scientific discovery in other spheres. More especially is this so where the importance of ecology has now come to be generally recognized. This, too, adds to the rich promise for future workers in the field the Haughley Experiment opened up.

As agriculture, in an age of enlightenment, becomes progressively aligned with the cultivation of health, as it ultimately must, not only will such experimentation be a necessity for understanding the role of the living soil for the health of plant and beast; it will, above all, be a necessity for understanding the nutritional process whereby man—short-circuiting entanglement with *dis*-ease—may come to live in the ease and order of health.

I.H.P

PART TWO

THE STORY OF THE HAUGHLEY EXPERIMENT

A Digest

Introduction

The Haughley Experiment was started in 1939 and continued until 1969. The present report, however, is concerned only with the post-war period, from the time in 1947 when the Soil Association became responsible for both its direction and financing.

In the following pages I have attempted to give a brief account of its purpose, and the programme that was adopted to carry it out. Next I have tried to present a digest of those observations and findings which, through the years, have pointed the direction which future fundamental research might usefully take.

For those who wish to study the findings in detail, full reference to the available records is given in the table of source-material which forms Appendix 12.

A short summary of how the experiment began has already been given in the Connecting Link. The scientific consultants responsible for its planning and direction, together with other key personnel concerned with administration throughout the twenty-odd years covered by this report, are given in Appendix 1.

Those Soil Association members, and others who, throughout the years, have given direct support to the Experiment through 'Friends of Haughley Research' or 'The Haughley Continuation Fund', received and studied the reports as they were issued and are, therefore, conversant with the purpose of the research programme and its progress.

Their interest in the present Digest will lie mainly in the chronological account of how, and why, the Experiment ended, and in the final conclusions to be drawn as to its value and achievements.

Such readers can obtain the necessary information covering these aspects by reading Chapters 5, 6 and 7.

Tables and figures have, as far as possible been kept out of the main text in the interests of readability, but the published summaries of the analytical work carried out in Dr. Milton's laboratory, from 1952 to

1964, together with his comments as Director of Research are given in full[1] in Appendices. All costs in the tables are in pre-decimal currency. Where immediate reference to any item therein is desirable, in order to help elucidate the main text, footnote references have been made to the relevant page.

Finally, I would like to acknowledge and record a profound debt of gratitude to my collaborator, Dr. R. F. Milton, and to Dr. Innes Pearse: without their advice and help my task would have been well nigh impossible.

1974 E.B.B.

[1] Except for those extracts already quoted in the main text.

I

The Case for Investigation

To enable this story of the Haughley Experiment to stand on its own, as a reasonably complete account, a certain amount of duplication of evidence given in Part I has been unavoidable in this chapter.

The field of interest, which gave rise to the Experiment, is that of *health*: health of the soil; of the crop; of the animal, and thus, by implication, of man.

It was planned to be a piece of fundamental research to explore the ecological interplay between these different links in the food chain, and their bearing on the health of each. It could be described as a search for conditions attendant upon biological wholeness. The founders[1] started with the assumption that apparently separate links in the nutrition cycle were functionally interrelated; interdependent, and inseparable.

They put it thus:

Nutrition is a basic biological function and requirement. It consists of a flow of materials and energy between organism and environment. This flow is a cycle: it passes through the interconnected forms of life—from the minute soil organisms to the plants, thence to animals and man, and back to the soil. This cycle must be studied as a whole, and these forms of life as part of it, and through successive generations, if we are to learn more about nutrition and health in each stage of the cycle.

The founders decided to study this biological cycle in operation, under three different systems of land use, on a farm scale. They derived much of their inspiration from some of the early pioneers in this field, who, as history is beginning to show, had visionary foresight when they insisted that nutrition is the basis of health, that soil fertility is the basis of nutrition, that soil–plant–animal–man is in

[1] See 'Connecting Link', p. 184.

effect, a single nutrition cycle, and that only when this cycle is studied as a functioning whole could any real advance be expected in the promotion of health and the prevention of disease, particularly in the metabolic and degenerative diseases such as cancer, heart disease, blood-pressure diseases, rheumatism, etc.

Special acknowledgement must be made to three of these pioneer-prophets. First, to the late Dr. Lionel Picton, general practitioner and for many years Honorary Secretary of the Cheshire Medical and Panel Committee, who were the authors of the 'Medical Testament'.[1] To requote two sentences from that now historic document: 'Our daily work brings us repeatedly to the same point—"this illness results from a lifetime of wrong nutrition".' 'For nutrition and the quality of food are the paramount factors in fitness. No health campaign can succeed unless the materials of which the bodies are built are sound. At present they are not.'

Second, to the late Sir Robert McCarrison, whose research work in India provided the first convincing experimental proof that the most important single factor in health is the kind of food consumed.[2]

Third, to the late Sir Albert Howard, whose life's work, and that of others based on it, produced a wealth of circumstantial evidence supporting his conclusions that the health-giving properties of even the right kind of food are dependent on the biological aspects of soil fertility, and to his prophetic hunch (for at that time there was no direct evidence of this) that the link would be found in the relationship between soil microflora and the quality of plant protein.[3]

The founders of the Haughley Experiment were deeply impressed by the importance of the observations, research work, and conclusions of these three men, and also by the practical results of putting their theories into practice, that were being reported from many parts of the world. The more they studied these reports, however, the more they became aware that there was a gap in the evidence concerning the part played by the soil-fertility factors.

The true nature of this gap, and the vital importance of filling it, can perhaps best be explained by first repeating one more quotation from the 'Medical Testament'. The passage occurs following a summary of McCarrison's findings. 'It is far from the purpose of this testament to advocate a particular diet. The Eskimos on flesh, liver,

[1] Published in 1939. See Part I, Chapter 2. See also Lionel J. Picton, *Thoughts on Feeding*, Faber & Faber, London, 1946.

[2] Cantor Lectures, see Bibliography.

[3] *An Agricultural Testament*, see Bibliography.

blubber and fish; the Hunza or Sikh, on wheaten chapatis, fruit, milk, sprouted legumes and a little meat; the islanders of Tristan da Cunha, on potatoes, seabirds' eggs, fish and cabbage; are equally healthy and free from disease. But there is some principle or quality in these diets which is absent from, or deficient in, the food of our people today. Our purpose is to point to this fact and to suggest the necessity of remedying the defect.

'*This at least may be said, that the food in all these diets is, for the most part, fresh from its source, little altered by preparation, and complete; and that, in the case of foods based on agriculture, the natural cycle is complete.*

'*Animal and vegetable waste–soil–plant–food–animal–man.*

'*No chemical or substitution stage intervenes.*'

The last few lines of this quotation, which have been printed here in italics, put forward no less than four possible causative health factors:

(a) Fresh food.

(b) 'Whole' food (all edible parts eaten).

(c) Food grown on soil to which all organic wastes are returned to complete the cycle.

(d) Food grown without chemicals or man-made substitutes for natural processes.

The most controversial factors, but not for that reason the least important, are the last two, which are directly connected with soil, and it was in connection with these that the gap in the evidence was apparent.

When these two factors are put into practice in modern systems of food production, they give rise to methods which have come to be known as 'organic farming and gardening'.

There seems no reason at all to doubt the constantly reported instances of health benefits claimed to result from eating organically grown produce. In the cases of human health, reports from individuals show that a change to organically grown (particularly compost grown) food has been followed by such things as the disappearance of rheumatism; the arresting of dental caries; greatly reduced susceptibility to colds and infections; curing of allergy symptoms; curing of supposedly incurable heart complaints, and so on. Reports from boarding schools show absenteeism from sickness reduced to an average as low as one day per year per child following a change to an organically grown diet; also improved dental health; improved standards of work and play; less quarrelling; virtually no colds or epidemics. In the case of special schools for backward, and in some cases delinquent children, near normal intelligence has been achieved, with the disappearance of anti-social tendencies. Too many well-authenticated reports of these various kinds are available to doubt them,[1] but when they are investigated, it is found that the change to organically grown food also involved, to a lesser or greater degree, a change from processed and refined foods to whole, fresh foods, and an increase in the proportion of them taken raw. There was therefore no means of determining how much of the benefits were due to these other factors, and how much to soil conditions.

In the cases of animal health, though the evidence was stronger, it was inconclusive; the gap was still there. The alleged results of a change to organic farming include the disappearance of deficiency diseases in livestock, including those only now beginning to be recognized as such, for example, contagious abortion and mastitis; the same output of animal products, whether meat, milk, wool or eggs, obtained with 10–15 per cent less weight of food; improved rearing results; better breeding records, and greatly increased longevity.[1]

Once again study of the evidence revealed that the change to organic methods was usually accompanied by the feeding of fresh, home-grown produce in place of the factory products, such as processed meals and oil cakes, previously used. Some reports, however, concerned animals reared mainly or wholly on pasture, without any supplements. The circumstantial evidence in these cases was certainly such as to call for further investigation.

The strongest evidence of the direct effect of soil conditions on nutrition was to be found in plants. The elimination of deficiency diseases, and the much greater resistance to pest damage, constantly reported as following a change to organic husbandry, were especially striking.[1]

Most challenging of all, to the orthodox views of the time, were the indications that resistance to pests, diseases, and to metabolic disturbances generally, increased with successive generations of organically nurtured plants and animals, thus suggesting that the effects of nutritional factors are cumulative.

Very striking evidence in support of this hypothesis is provided by the famous 'cat' experiments of the late Dr. Pottinger[2] and the equally

[1] See Part I, Chapter 6.
[2] See Appendix 11.

cogent evidence resulting from the investigations of the late Dr. Weston Price.[1]

As a result of close study and sifting (and some personal testing) of many of the reports referred to above, it became clear to the founders of the Haughley Experiment that a type of comparative research different from any existing agricultural research was urgently called for. They began to suspect that all disease might be a symptom of unbalance between a living organism and its total environment, and that the key to health would not be found through the fragmentary approach of seeking the cause of specific diseases, but in studying living function between organisms and their environment as a dynamic whole.

This view was strongly enforced by the results of the Peckham Experiment, which were made available at about the same time as the Haughley Experiment was being planned.

When, in 1926, Doctors Scott-Williamson and Innes Pearse (already widely experienced research workers in early diagnosis and investigation of disease) established the Pioneer Health Centre at Peckham (the family club which became the first laboratory in the world for the study of human health) 'they already had come to the conclusion that no headway would be made in the understanding of health by researches into the nature of disease.'

They had good reason, from their own personal experience, to observe that in spite of rapidly advancing medical science and vast sickness services, disease was not being stamped out. Even the development of so-called preventive medicine seemed largely to be accompanied by new diseases cropping up to replace those that have been banished. They deduced, as a result of their researches, that wherever 'positive' health was encountered it was seen to be an active *process*; something quite different from mere absence of disease. They even went so far as to state their belief that health could be as infectious as disease, and could also grow and spread. They further sensed that 'health had its own *pattern of behaviour*—a pattern quite different from the pattern of disease', which would become apparent given suitable environmental circumstances procurable by experimental methods. It was this that the Peckham Experiment was designed to study.

Small wonder, then, that in due course these two medical biologists proved to be among the most valued consultants to advise those planning the Haughley Experiment, for the latter was a parallel

[1] See Bibliography.

study. It, too, was designed to disclose the pattern of behaviour evinced by health in living soil, and in the living organisms dependent on it.

The following quotation admirably sums up the case for investigation:

'Disease has come under the microscope . . . our lenses grow more powerful and our technical ability upsurges, but all this study of disease does not reveal to us the laws of health. It is *health* itself which must be studied. We must devise laboratories where we can put health, too, under a lens, look at it, discover how it behaves, and find out in what conditions it can grow and spread.'[1]

For health, whether of man, beast, plant, or soil, is much more than the absence of disease. This we shall see as we proceed.

[1] I. H. Pearse, *Health, of the Individual, of the Family, of Society* (Pioneer Health Centre, 1971).

2

Planning the Framework for the Experiment

It was decided early on that any unit of study must be large enough to enable all the normall interrelationships to have full play, and that such a study would have to continue through many successive generations of plants and animals.

Research into 'wholes' is notoriously difficult. The founders decided as a first step to initiate a long-term experiment, on a full farming scale, to compare and contrast the nutritional effects, on successive generations of farm animals, of food grown from successive generations of crops, nurtured under different systems of soil treatment, on adjoining units of land of similar soil types, and under the same management.

Almost from the start it became apparent that such a piece of research work could be established only if it was privately financed,[1] for the concept on which it was based forced upon it techniques at that time too revolutionary to be accepted in official circles. In the first place it involved study of the interrelationships of soil, crops, and grazing animals in a fully rotational farm system, which entailed scrapping altogether the conventional small randomized plot method of ordinary agricultural research. In the second place, it also involved, to a very large extent, abandoning any hope of expressing the results statistically, since an attempt to eliminate any of the variables inherent in biological function in a farming 'organism' would destroy the 'whole' it was intended to study. Since all science has for so long been the slave of statistics, these departures from the conventional inevitably led to an assumption among agricultural scientists that the Haughley Experiment was scientifically valueless—a view that is now changing. Finally, an ecological search for quality in food, as comprehensive as that envisaged, was not purely agricultural. Its objective was to enlarge our knowledge of the nature of health itself.

[1] For method of financing the experiment see Chapter 5. For management personnel and scientific consultants see Appendix 1.

Most agricultural research tends to be mainly concerned with techniques for increasing quantity production without taking into consideration the effects which such techniques have in disrupting the biological cycle in soil–plant–animal–man: disruptions which could have a profound effect on factors upon which the health and fertility of the components of the biological cycle depend. Medical research, as already pointed out, tends to concentrate on causation of specific diseases, and ignore conditions requisite for health or wholeness as a positive development process.

The experiment envisaged, thus fell between two stools.

The Ministry of Agriculture considered it to be a matter for the Medical Research Council, while the latter regarded it as the concern of agriculture.

At that time even in the science of pathology, the idea of such a development as agro-medical research was not taken seriously, whilst as for agriculture making a positive contribution to health, this had barely been thought of at all.[1]

The site chosen for the experiment was two adjoining farms in the parish of Haughley, near Stowmarket in Suffolk, that were already being farmed as one holding.[2]

The area, comprising 216 acres, was divided into three units to be run as three self-contained small farms. The divisions were carefully chosen so that such slight variation in soil type, aspect, etc., as existed should be represented in all three. The fairness of the split was subsequently confirmed by a soil survey undertaken by the Rothamsted Agricultural Experimental Station.

Two of the three sections, each having a working acreage of approximately 75 acres, were destined to be the stock-bearing sections, each to carry its own self-contained herds and flocks of *not less than three species of farm livestock.*[3] The third section of only 32 working acres was to be a stockless section.

[1] Except for a few pioneers, such as the late Professor Lindsay Robb, who constantly argued the desirability of a single Ministry of Health and Land-Use (see 'Medicine and Agriculture', *Guy's Hospital Gazette*, 20 July 1968), and the late Dr. G. Scott-Williamson who already had observed, with regard to disease, that 'Causes lie in the environment—consequences are seen in the patient.'

In 1935 he had already attached a nutritional farm as an essential part of his practical experiment into the nature of health, the Pioneer Health Centre, Peckham. By 1957 this view was beginning to be accepted in some medical circles (see Postscript).

[2] For more detailed description see p. 184.

[3] An ecological approach to soil fertility precludes mono-culture of livestock equally with mono-culture of crops. Maximum diversity within practical possibilities is the objective.

All three were to be established with seed, and two with foundation animals, of common origin and thereafter each was to grow and breed its own requirements.

One stock-bearing section was to be called the Organic section, the fields of which were to receive only the crop residues and animal manure produced on the section.[1]

The fields of the second stock-bearing section were likewise to receive the crop residues and manure produced on it, but also to have standard supplementary applications of chemical fertilizers. Because it received both, it was to be called the Mixed section.

The fields of the Stockless section, which, as the name implies, carried no animals would receive the crop residues it produced, together with chemical fertilizers, but no animal manure.

Only the creation of these three divisions had been completed when World War II broke out.

After the war, planning began in earnest—the challenge of attempting to explore 'wholeness' having been accepted, it became necessary to devise a framework in collaboration with consultant scientists, which would allow the effects of the biological and ecological 'unknowns' to be observed and recorded.

In a wilderness area, where no human exploitation exists, it is well known to naturalists that, assuming the area to be big enough, nature establishes a biological balance. With the result that although the many species of plants and animals comprising such an environment subsist by 'preying' on each other, almost never does any species become extinct, nor does it build up to pest proportions, so that the whole ecological system of soil, plants, insects, birds and animals (herbivora, omnivora and carnivora) continues to function and prosper.

The planners asked themselves—to what extent can nature and man, working together in harmony, establish a similar biological balance among the living organisms involved in an agricultural food chain, which, by its very nature must be subject to a large degree of human interference.

The rules laid down for the operation of the Organic section of the Haughley Experiment were designed in the hopes of getting an answer to this question, and also to serve as a control against which to learn more of the effects on the biological cycle of the man-made additives which the Mixed section received, as well as the effect of the short-circuit imposed on the cycle in the Stockless section through depriving it of animal residues (other than local wild life).

[1] See paragraph on 'closed cycle', p. 204.

The most important of these rules were:

(a) That all three sections should adhere to a strict rotation. On the two stock-bearing sections this to be identical and based on a ley system of farming.

(b) No importation of feeding stuffs, or pulse and cereal seed grain, to be permitted on either stock-bearing section, whether from outside or as a transfer from one section to another. It was recognized that a possible exception to this might have to be small amounts of mineral food supplements.

(c) On the Organic section it was laid down that as nearly as possible to a closed cycle must be operated. Only livestock end-products (milk, eggs, wool and surplus animals and birds) to be exported, and only breeding males, some small seeds (see below) and an organic mineral food supplement of known origin, and composition—dried seaweed—to be imported.[1]

(d) Finally, the unit of time, for any comparisons or assessments, to be a *full rotation period.*

As the work developed, and these rules were put into practice, they met with periodic criticism, due largely to misunderstanding.

Three aspects in particular, many visitors to the farm, and even some keen supporters seemed to find great difficulty in understanding.

One was the possibility that, with both crops and livestock, *functional action* in a 'whole' (i.e. an ecological situation) may differ from *operational behaviour* in fragmented situations.

Another was the way in which long-term basic research (the only kind that can hope to reveal *how* nature's ecological principles work) had any *practical* value. They failed to see that unless applied research, in any biological field (being, as it is, under constant pressures to find quick answers), has its roots in more fundamental research, it can, and often does, lead to practices that have disastrous long-term effects.[2]

But what some critics found hardest of all to understand—and this applied especially to farmers and agricultural technologists—was why basic research into 'quality' required a closed cycle on the organic section.

Their difficulty over this derived from a fundamental misconception concerning the nature of the experiment.

To people whose whole training was based (and indeed was de-

[1] Later small quantities of fish meal were allowed for the poultry: a concession, on economic grounds, to the critics (see paragraph 2 on p. 205).

[2] Vide chlorinated hydrocarbon pesticides, to mention only one.

pendent) upon quantitative measurements, the need for a new framework and a new yardstick with which to investigate 'quality' was, apparently, incomprehensible. The result was their complete failure to appreciate the significance of the closed cycle in any observational study of ecological function between organism and environment.

As will be seen in due course, the not inconsiderable achievements of the Experiment *resulted* from the closed-cycle concept, and would not have occurred if the arguments of the critics had been accepted.

This failure to understand the incompatability between fundamental biological research (concerned with exploration of unknown ecological factors) on the one hand, and applied research and demonstration (concerned with economic aspects and bulk yields) on the other, proved to be widespread and persistent in certain influential quarters, and this to some extent bedevilled the Experiment throughout its life, ultimately bringing it to an end.

Because of these confusions of thought, and in the hope of saving the reader from similar misunderstandings, we give below, in question and answer form, the scientific reasons for incorporating the rules, listed above, into the framework of the Experiment.

In so doing, we make no apology for a degree of repetition, where this occurs.

Why a Strict Rotational System?

Rotational systems are man's attempt to apply to farming, lessons learnt from observation of natural processes. There is always diversity in a 'natural' environment,[1] with harmony depending on the balance between competition and co-operation among its component parts, which adds up to complex patterns of mutual interdependence.

By rotating the crops, and livestock, on a farm (each of which when grown by itself tends to encourage the proliferation of particular soil microflora, insects and small fauna) diversity of the soil population as a whole is fostered. Also, which is probably equally important, the accumulation of any one type of specific metabolic product is prevented.

[1] One of the dangers in many modern techniques, especially the use of biocides but also even the consistant introduction of hybrid varieties, is that they diminish diversity in the environment.

Why a Ley System?

First, because the period under the temporary pasture—especially when, as in the case of the Organic section, such pasture consists of a mixture of deep-rooting grasses, legumes and herbs—is the time of rehabilitating reserves, which may be utilized by the subsequent arable crops. It is during this time that the humus reserves are increased, the nitrogen fixation accelerated, and the soil organisms encouraged.

Secondly, because ley farming is the most efficient way to enable each species of farm livestock to bring, via direct grazing, its own influence to bear, on every field, once in each rotation period.

Why No Brought-in Foodstuffs or Seed Grain?

The answer to this is part and parcel of the answer to the next question:

Why a 'Closed' Cycle?

Because the particular 'behaviour pattern' which it was the purpose of the experiment to endeavour to 'observe' and study was the dynamic nutrition flow throughout an agricultural food chain of soil–crop–animal and back to the soil again, in a particular place, functioning as a whole, for several revolutions of the cycle, through successive generations of all its parts. Only components of such a closed system can be of known history, and they may—as they pass from one organism to another—take with them this past history and bestow the consequences thereof to the new organism. Such a possibility is well worth fundamental study since it may give us a glimpse into many of the factors which we associate with quality. Such factors may be concerned with natural insusceptibility to disease, particularly those of a virus[1] nature which have their origin in the protein metabolism of the cell. If, however, we bring into this system from outside, components of unknown history or of different history from that within the closed cycle, then we are introducing substances of alien specific patterns; complications which must confuse any interpretation of interplay between organism and environment and which may profoundly affect the study of quality. It follows, therefore, that the least amount of introductions of this kind the better—if we wish to

[1] Also relevant is recent work on chromosome-enzyme action.

understand the implications of our experiment. We must limit such introductions to those likely to have the least effect.

The ley-rotational farming system laid down allows of the export of only milk, eggs and wool and the carcasses of the sold-off animals and flocks. In terms of removal from the soil, this works out to infinitesimal amounts of mineral substances (at the most $\frac{1}{500}$ of the reserves of the top 9 inches of soil each year).[1] Since an organic rotational ley system incorporates deep-rooting herbs in the pasture which continually bring up nutrients from the deeper layers, and nitrogen fixation is more than in balance, we may consider the losses from the farm as negligible. For this reason it should not be necessary to import nutrient substances or their precursors into the biological cycle. In fact, we should be in a position to arrange and increase our total fertility so that what we export is surplus to our requirements.

It is not surprising that some people found this difficult to accept, because it is heresy to orthodox teaching, which declares that whatever is taken out of the soil in crops and stock must be replaced, pound for pound, if loss of fertility is to be avoided; from which it was deduced that on the Organic section of the Haughley Experiment a rundown is inevitable where an almost closed cycle is operating inwards, but where there are substantial leaks outwards (sale of milk, eggs, young stock, etc.) The Experiment ended before this deduction was proved right or wrong,[2] but the theory behind it is less self-evident than appears at first sight.

In the first place it ignores the extent of the contribution to the cycle made by the atmosphere. It also ignores the history of evolution, as usually taught, for this indicates unequivocally that when *everything* goes back to the soil, i.e. a true closed cycle, a steady enrichment of the environment takes place, with *increasing* soil fertility, so that each new generation of 'return' enables richer, more complex life-forms to emerge (from algae and bacteria to rain-forest and elephants as it were).

From this it follows that there must be a point at which a *stabilized* level of soil fertility can be maintained indefinitely, *short* of a 100 per cent return. Finding out what that point is, seems the best way to establish a real 'control' for evaluating the various methods of soil manipulation as practised in agriculture.

Further than this, available evidence increasingly supports the view that the complexities of the living world of the soil and plant

[1] See Chapter 7, p. 276. Also Appendix 5, p. 330.
[2] See Chapter 5.

cannot be fully interpreted in terms of simple arithmetical sums of addition and subtraction of materials. 'Forces', of which we as yet understand very little, enter into it, and may even prove to be the *most* important factor in soil fertility. The proved growth-response of plants to electromagnetic radiations, a response greatly exceeding that from nutrient applications, lends support to this view. It is not improbable that the next great surge forward in biological and agricultural knowledge will be in this field.[1]

In the meantime the possibility of operating a 'closed' biological cycle which actually *increases* the potential fertility of the soil, was accepted as a working hypothesis, since such a cycle, correctly operated, *could* bring into the system more available minerals from denuded rocks, clays, etc.; more nitrogen by fixation by aerobic bacteria, and more humus by utilization of the carbonates in the soil and the carbon dioxide in the air than are removed in produce. The catalysts in this process are sun and water, together with our own control or encouragement of the appropriate biological energizers, i.e. the vast living population of the soil itself, which in the aggregate can amount, in fertile soil, to the astonishing weight of over 10 tons to the acre, and the by-products of whose activity, according to the late Sir John Russell[2] are *all* that the human cultivator of the soil can hope to garner!

The type of (almost) closed-cycle farming experiment which has been described does not allow the export of so-called cash crops except on the Stockless section. All crops grown on the other sections (whether the Organic or Mixed) must be consumed by the livestock belonging to that section (either as food or bedding) and all final end-products, manure and crop residues, returned to the soil.[3]

If we accept the experimental cycle as defined above, with exports confined to livestock products, we must clearly define the type and maximum amount of substances which can be *imported*. Where do we draw the line? Clearly feeding stuffs and plant and animal residues

[1] See Peter Tompkins and Christopher Bird, *The Secret Life of Plants*, Allen Lane, London, 1974; also C. L. Kervran, *Biological Transmutations*, Crosby Lockwood Staples, St. Albans, 1972.

[2] See Sir John Russell, *World of the Soil*, p. 137.

[3] It should be emphasized, however, that this is only one type of closed biological experiment. An extension would be to bring man into the cycle. Then the produce (dairy and 'cash crops') would be consumed by this human community, whose waste products in true bionomic economy would in turn be brought into the cycle. In the past, a true peasant economy must have operated just this kind of closed-cycle.

cannot be imported, but, in the area available we cannot avoid bringing in the grass seeds for the leys, and other small seeds such as brassicas, beets, etc. For choice, these should be the only importations on the organic section.

The value of the other two sections is in the study of the change in effect produced by introducing 'man-made' additives in mass concentration to an otherwise closed biological cycle. The Mixed section can be regarded in this light and considered as a compromise between a rotational return system and an attempt at forcing by applying high concentrations of readily soluble nutrients. On the Stockless section, the rotational experiment may be followed but with a deliberate 'short-circuiting' of the biological cycle, by the omission of a return of animal wastes, and by the omission of the type of broken-down plant residues produced by a compost heap or dunghill. In order to furnish valuable information, rigid control of the introduced material must be followed, which entails confining such introductions to inorganic materials.

In general it should be stated that this approach leaves a wide latitude in scope and procedure, for differing farming techniques. Particular techniques which might be applied at any given moment could be such as profoundly to affect soil aeration for example, or other factors such as drainage with concomitant transferred effect throughout the biological components of the closed system. When such techniques resulted in adverse effects, which in commercial terms might be considered as farming failure, they could nevertheless be of equal value—contributing to the wealth of our knowledge of the processes involved—as might changes of technique which result in high-crop yields.

Why the Rotation Period as the Unit of Time?

There are several reasons for this.

(a) Every crop leaves an imprint in the field that grows it, and this has some effect on the subsequent crop. In certain cases, such as temporary pastures, the effect may extend to a succession of following crops.

(b) A similar mutual imprint arises between a pasture and the animals that graze it. Both the grazing management and the different livestock species play an important part here.

(c) On a self-contained farm, when farmyard manure from animals housed in the winter is applied to a field producing little of the food and litter of which it is composed, a transference of fertility takes place from the producing to the receiving fields.

Thus, where different systems of land use are being compared, it is only the results of the full rotational cycle that will bear comparison. When every field has grown every crop, and has gone through its full period under pasture, on which all the different kinds of stock have grazed, then the total production over that whole period, of any particular crop in one section, can be compared, quantitatively and qualitatively, with a similar crop total in another section. Year-by-year comparisons have very little significance, partly on account of the factors mentioned above, and partly because seasonal differences, accidental differences, and inherent differences in the fertility of different fields, can produce totally misleading data from which false deductions can be drawn. When the rotation period is taken as the unit of time, however, all these differences tend to be ironed out.

The same principles apply in the case of the Stockless section, but here a five-year rotation is in operation, so the unit of time is five years. That is to say the cycle revolves twice in the Stockless section for every once in the other two sections.

This, then, was the way in which the Experiment was set up in 1947–8.[1]

[1] For details of stocking and cropping see *The First Twenty-Five Years* and subsequent annual reports at the Soil Association Headquarters. Also Appendix 2.

3
Early Findings

By 1952 the rotations had got under way, and the purchases of foundation livestock had been completed.[1] Two closely related dairy herds of Guernsey cattle had been established (one on each of the stock-bearing sections). Likewise two flocks of light Sussex poultry. The third species of livestock at that time was pigs, home-bred for many generations. In 1960 Clun Forest sheep were substituted because sheep make less demand on grain crops than pigs, and are better scavengers of surplus fodder.

The chosen rotation on the two livestock sections, from 1952 to 1959, was oats; barley; pulse; wheat; arable silage (oats and peas or oats and tares) undersown with a four-year ley mixture, leading to four years of temporary pasture, used for grazing and hay, then back to oats. In 1960 this nine-year rotation was slightly modified as follows: wheat; roots and forage; barley; pulse; oats, arable silage with undersown ley, as before, but returning to wheat on plough-up instead of oats. This, now ten-year rotation provided (per stock-bearing section), in any one year, approximately 20–25 acres in temporary leys, and 40–50 acres in arable crops. In addition, roughly 6–8 acres of Lucerne was grown (outside the rotation) and there was about 7 acres of permanent pasture. These figures, of course, applied to both sections.

On the Stockless section, of 32 working acres, a five-year rotation was established of four arable cash crops followed by clover ploughed in[2] (in a suitable year a crop of seed was first taken, and the aftermath ploughed in).

By 1952 then, the scene was set, and the true preparatory stage—namely the first complete rotation period—was ready to start, with all the complicated record-keeping which the experiment now entailed.

[1] For more details, and methods of establishing the herds and flocks see Appendix 2.

[2] For details and varieties see Appendix 2.

Dr. R. F. Milton[1] was invited to plan and direct a scientific programme of measurements, analyses and assessments, to run parallel to, and be correlated with the farming observations.

It was not expected that the first unit of time, i.e. the initial rotation period (1952–61), would produce significant results, or even much in the way of information, for it was recognized that it would take this period for full sectional integration, between soil, crops and stock, to take place. In the event, however, it yielded a surprisingly rich harvest of valuable pointers.

As a consequence it was considered worthwhile in 1962 to issue a first detailed report[2] for the benefit of members of the Soil Association (the sponsoring body since 1947; see pp. 295, 370).

The rest of this chapter consists, mainly, of condensed extracts, with comment, taken from that report.

We start with part of the introduction, contributed by Dr. Milton, to the tables of figures summarizing the first ten years of his analytical work on the soil.[3]

'When I was asked to participate in the Haughley Experiment, I was already acquainted with its size and scope, and of the criticism that the experiment could not possibly produce conclusive results in that it was not susceptible to statistical treatment, but my experience in the wider biological field had taught me also, the limitations of the statistical approach. The statistical formula is valid only if the number of variables is known and limited, and in any experiment involving a biological organism the variables can never be assessed—let alone controlled. In most other agricultural research stations, replicated plot experiments are carried out and the results statistically treated. Such results may give valuable information, but application of the results outside the limitations of the plot experiment may be quite unjustified, and only too often such is the case.

'Now the Haughley Experiment is concerned with differing methods of farm management and it involves the complicated interplay and balance between soil, crop and animal. The limits are boundless and cannot be anticipated or controlled. The experiment, therefore, is to be considered as a pioneer or proving venture which may reveal effects upon soil, plant and animal that could never have been obtained by plot or controlled agricultural experiment.

[1] See p. 295.

[2] *The First Twenty-Five Years, 1938–62* published by the Soil Association. See p. 370. See also Appendix 3.

[3] For more details, with graphs, the reader is again referred to *The First Twenty-Five Years*, also to Appendix 3.

SOIL

'Within this framework then, it became necessary to elaborate an investigational programme designed to obtain maximum information concerning relationships between soil, plant and animal. To this end, monthly soil samples have been taken for examination of levels of available minerals and other recognized indications of "fertility".

'We have pursued the idea that variations in farm management experienced on the three sections would ultimately be manifest in changes in soil structure. Certainly the field highest in humus has a "fluffiness" which can be embarrassing when preparing a seed bed. The fields on the Stockless section liberally treated with chemical fertilizer undoubtedly tend to walk heavily—the top soil becomes dusty quickly and the character changes readily after a rainy period. But we have been unable, so far, to obtain definite measurements which back up the contention that the soil structure is being lost on the Stockless section. Measurements of aggregates show greater variation over the individual fields than we find on the different sections.

'Considerable work has been carried out in the elaboration of satisfactory methods for soil differentiation, and the only method at all found likely to bear fruitful results is that of penetrometry. Many thousands of penetrometer determinations have been carried out during the past few years, and providing that one relates the findings to the prevailing conditions at the time of measurement, then comparison between the sections indicates clearly the increased tendency of the fields on the Stockless section to become panned when assessed against the corresponding arable fields on the stockbearing sections. This is associated with the drop in humus content on the Stockless fields which has occurred during the period of observation—despite the introduction of a fallow and ploughed-in green crop once in the five-year rotation. The tendency towards a drop in humus content on the Stockless fields is not such a marked annual feature as was observed in the early days of the investigational work, and it is felt that, as the level approaches 2 per cent humus content (the lowest figure compatible with reasonable crop growth according to one authority), then the rate of fall will tend to decrease. On the Mixed and Organic sections the humus levels are being maintained or raised slightly—but this varies according to the individual fields some of which require considerable attention in order to maintain fertility. It is an observed fact that conditions of aeration and drainage are dominant factors in these cases. In such fields any improvement in fertility has always been associated with an increase in the humus content of the soils.

Water-holding Capacity

'We have continued to make observations on the water-holding capacity of the soils of all the fields on the farms, and confirm the earlier findings of correlation between this factor, the humus content, the degree of aeration, and the behaviour of the land in times of drought and flood.

pH

'We have, consistently, throughout the whole period of years found all the fields to be remarkably highly alkaline, with considerable variations in the degree thereof throughout the year. The changes do not appear to follow a consistent pattern, but the greatest variations have always been found on the Organic section. The pH levels at the most alkaline points would—according to theoretical considerations—render the soil deficient in certain trace minerals due to unavailability of the insoluble phosphates and carbonates. In point of fact, however, there have been no indications of this either in the appearance of the crop or according to the analytical findings.

Available Minerals

'Citric soluble phosphate, potash, soda and nitrate, have been determined monthly for some years (in the case of the two former throughout the whole of the nine-year period).

PHOSPHATE

'We have found marked variations from month to month in citric soluble phosphate levels. On the Stockless section the variations are irregular, and apart from a generalized rise in November and December on those fields which have been subjected to fertilizer treatment in the autumn, no general pattern is observed. With fields on the Mixed and Organic sections generally, a peak was apparent in the later summer months. This is the general pattern of rise and fall; it suggests a correlation between enhanced biological activity in the soil and most rapid plant growth.

'We have confirmed the observation that the largest fluctuations occur in fields with highest humus content, particularly if growing arable crops.

CITRIC SOLUBLE POTASH

'Monthly variations in potash levels in the soil are a characteristic feature. By and large the variations follow the pattern described above for citric soluble phosphate. Greatest variations occur in fields with highest humus and the peaks are in the growing months. Least variations are in fields under leys and are greatest with arable crops. Some fields on the Stockless section show high levels in the winter (following addition of fertilizer), and fairly low results in spring and summer.

NITRATE

'Variations in nitrate levels are also found in the monthly soil samples. The pattern is not dissimilar to that shown by phosphate and potash. Incubation of the soils has shown that remarkable increases in nitrate concentration will result, and again there appears to be a correlation between humus content and potential nitrate production.

'The fields on the Organic section rich in humus show also increasing levels of total nitrogen (nitrogen bank). This rise is not apparent on the Mixed and Stockless sections despite the yearly additions of nitrogenous fertilizers. The increase, therefore, must arise as a result of a stimulation of nitrogen-fixing organisms. The inescapable conclusion is that high humus content and enhanced microbiological activity are interconnected.

'It is worth noting that the methods of determination of nitrate in soils, which had to be specially developed in connection with this work (*Analyst*, February 1955) is now widely used by soil analysts.

Trace Element Determination (Minor Elements)

'We have also had to devise original methods for trace-element determination (copper, cobalt, nickel, zinc),[1] based on chromatographic techniques, and these methods have been published in recognized scientific journals (*Analyst*, September 1958 and July 1959) and applied in fields other than soil analysis. In particular we have devised procedures for the assessment of available trace mineral levels in soils,

[1] In view of the alleged importance of magnesium in plant and animal nutrition our findings for soil Mg over a period of years are given in Appendix 7, p. 348.

and we have carried out in recent years monthly determinations for these elements on all the fields on all the sections.

'The pattern of seasonal variation with some of the minor elements (manganese, molybdenum and zinc) is very apparent (see Appendix 3), but with copper, cobalt, and nickel, this change has not been noticeable. Levels of available trace metals are adequate on all fields and there is no suggestion of sectional differences or of loss of availability due to changes in alkalinity.

'Much has emerged from the soil analysis carried out on the samples taken from the Haughley fields during the past ten years, the results of which are of fundamental importance in furtherance of the understanding of the chemical and biological processes concerned with plant growth. In a fertile soil there is an undoubted symbiosis between the soil organisms and the growing plant, and our results have highlighted this biological fact. Factors which influence microfloral distribution (e.g. poor drainage and lack of aeration) and soil temperature, considerably influence the growing plant, and this has been manifest in the availability of the plant nutrients in the soil. It would seem established that soil micro-organisms play a predominant part in the release of so-called bound minerals, so as to render amounts available which can be assimilated by the plant as required. The addition of chemical fertilizers will influence this mechanism (if only due to the mass action law of reversability of chemical reactions) and there is abundant evidence by other workers to confirm this fact. (See Sir J. Russell, *World of the Soil*, p. 64.) Our work has further shown up the futility of assessing the requirements for a crop from a single analysis of the soil taken at any time during the year—a practice which is now in disrepute.[1] It has also shown the importance of building up a high humus content in the soil and the influence thereof upon soil structure and fertility. Farmers who take cognizance of these facts are planning for the future as well as reaping the immediate harvest. We feel that there is little more to be gained by continuous monthly observations of this nature, now that most of the fields have completed a full rotation. We intend, however, to make periodical analyses of the soils so as to ascertain the trends and to develop further tests for assessment of soil structure changes.'

In addition to the soil work, Dr. Milton carried out comprehensive analyses of the farm produce throughout the period. This included livestock products as well as crops. Summaries of the findings will be found in Appendix 3.

[1] See Graph 1 on p. 217.

1952 MONTHLY SOIL ANALYSIS

on a field of Barley in each Section of

Haughley Research Farms Ltd. - Soil Association Ltd.

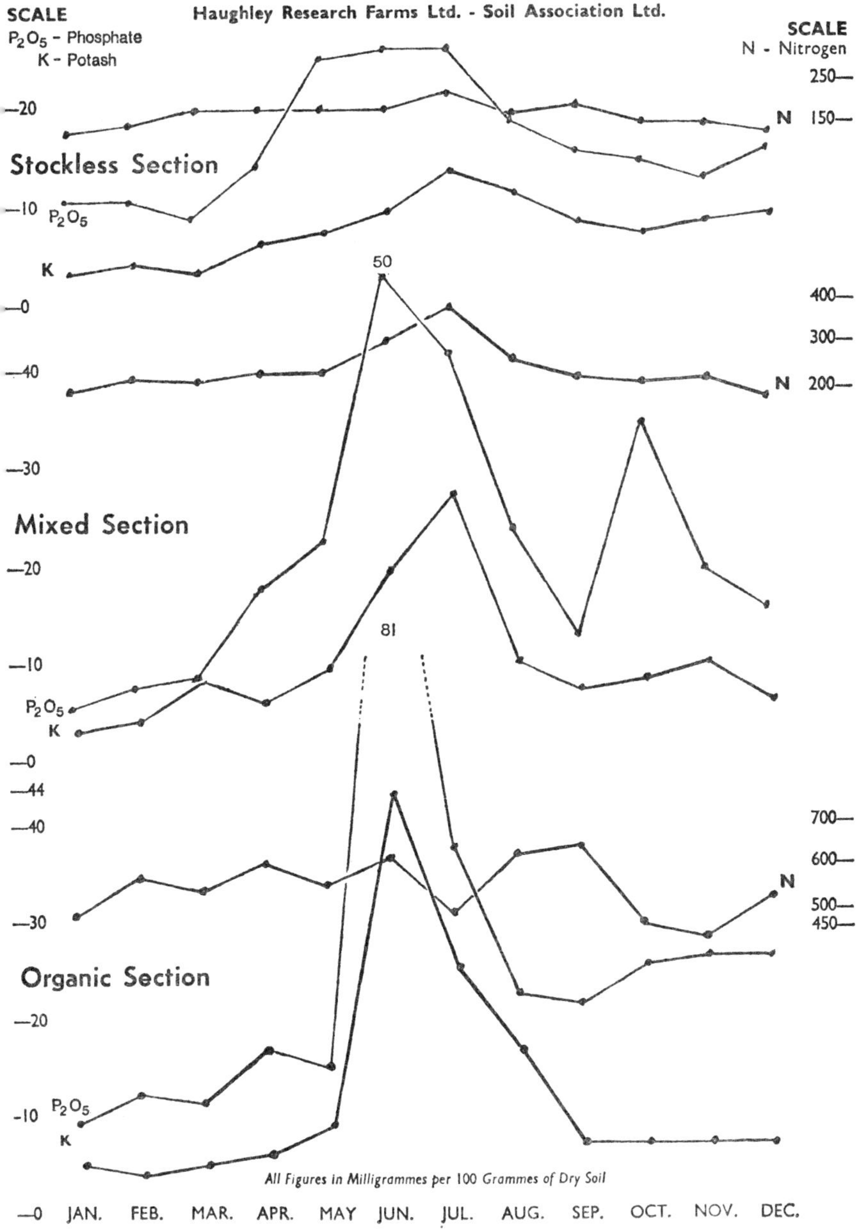

Bulk yields of the principal crops during the period are shown in Table I below.

The seed for each type of cereal or pulse was initially the same stock on all three sections—thereafter each crop was grown from seed saved from the previous year's crop; each from its own section.

The yields shown in Table I below are all well below those normally achieved today (1972) under modern intensive methods and with new seed varieties, but they were well up to national averages for the district at the time. Their relevance, however, lies in (a) the comparison of the ten-year averages on the three sections, and

TABLE I

Average bulk yields of crops, expressed in cwt. per acre per year, for the years shown in column 2, are given for the three sections in columns 3, 4 and 5. Figures in brackets are the actual yields for the last year (1961) from seed mostly tenth generation, or more.

	Columns 3, 4, 5, represent average yearly yields per acre for the years given below. [1961 yields in brackets]	*Cwt. per acre*	*Cwt. per acre*	*Cwt. per acre*	*Average annual cost per acre (in shillings of fertilizers applied)*		
CROP	YEARS	ORGANIC	MIXED	STOCKLESS	O	M	S
Wheat	1952–61	22 (21)	26 (24)	24 (18)	Nil	55/–	78/–
Winter oats	1953–61	23 (18)	24 (20)	Not grown	Nil	52/–[1]	—
Spring barley (Spratt Archer)	1952–5	20 (20)	21 (26)	22 (28)	Nil	54/–	58/–
Spring barley (Rika)	1955–61	28 (22)	27 (26)	32 (32)	Nil	44/–	50/–
Winter beans	1952–61	11·5 (16)	13·8 (21)	15·9 (22)	Nil	41/–	41/–
Peas (Dun)	1952–61	10 (9)	7 (8)	Not grown	Nil	18/–	—
Arable silage (Green weight)	1959–61	118 —	135 —	Not grown	Nil	55/–	—
Ley hay	1955–60	35 —	41 —	Not grown	Nil	46/–	—
Ley hay	1961	(36) (48)	(35)	Not grown			

[1] = Average for 1960 and 1961 only. Before that winter oats followed ley, and did not receive fertilizer.

(b) (probably more significant) comparison of the ten-year averages with the actual yields for the last year of that period.

The first published report, referred to on p. 212, which covers the above period (1952–61) included the following note on seed-dressing usage:

'On the Mixed and Stockless sections, mercuric seed dressings are used on all cereal and pulse crops, destined for ripening. No seed dressings have been used on the Organic section, except in certain years on wheat, starting with the first three generations. Then, after six generations without dressings, the crop was affected by bunt in a particularly bad season for fungus attack. Since then, part of the wheat crop has had the seed dressed with formaldehyde and part has been heat treated, in order to preserve the continuity of an undressed strain. Chemical dressings will be dropped again as soon as possible.'

It will be noted from the table that in almost every case a slightly higher yield was obtained where fertilizers were used (though not as much higher as might have been expected). But when the averages are compared with the figures for the last of the ten years it will be seen that there is remarkably little difference in the *pattern* of production between the sections.

In other words Table 1 shows that yields on the Organic section did not decline either in total production, or in proportion to the mixed; and this after eighteen years (in some cases more) *without any fertilizers at all.*

It may be recalled that one of the questions the experiment was designed to answer was 'Can fertility be sustained consistently without the addition of artificial fertilizers?' It was not concerned with finding out the maximum yield obtainable in any one season by any one method.

The whole of the produce of both the Organic and Mixed sections was converted, via the 'animals', into livestock products. Because of the somewhat lower bulk yields on the Organic section the quantity fed on that section had to be rationed more severely than was the case on the Mixed section.

For example, the allowance of concentrates (all home grown of course) for the milking cows in the winter of 1960–1 is fairly typical. The figures represent lb. fed per cow, per day, per gallon of milk produced.

	lb. per day	
	Organic	*Mixed*
1st gallon	1	1½
2nd gallon	2	2½
3rd (and each subsequent) gallon	3	3½

Thus a mixed cow giving 4 gallons per day would receive 2 lb. per day more concentrates than one giving the same amount on the organic section. In the case of higher yielding cows the gap was still greater because a ceiling of about 10 lb. per head per day was imposed on the organic herd.

Similar differences occurred with fodder, rations for which, in that same winter, were in lb. per head per day.

	Organic	*Mixed*
Hay	11	11
Silage	40	Ad lib
Straw	5½	—

Exact quantities and composition of rations varied from year to year as they depended on available supplies, i.e. each year's harvest results. For example, 1961 was a good year for fodder crops, so the winter rations for 1961–2 (again in lb. per head per day) were increased to:

	Organic	*Mixed*
Hay	11–12	15
Silage	50	Ad lib
Straw	5½	5½

This was sufficient to cover maintenance *and* the first gallon of milk, so the concentrate rations that winter were:

	lb. per head per day per gal.	
	Organic	*Mixed*
1st gallon	—	—
2nd gallon	2	3½
3rd (and each subsequent) gallon	3 (up to a total of about 10 lb.)	3½ (no upper limit)

(Rations for other livestock, and composition and analyses of all rations are given in Appendices 2 and 3.)

Perhaps the most surprising finding of the first rotation period was how little obvious correlation there was between the above food intake figures in the two herds and their output in terms of milk production.

Figures for total production for the seven years, 1955–61 showed:

	lb. Organic	lb. Mixed	lb. M difference
1955	67,646	68,375	+ 729
1961	75,872	73,776	—2,096
Average for the seven years	75,605	70,510	—5,095

(For more details see Appendix 3.)

It will be noted that in the first full year of milk records (1955), the herds being then mainly foundation stock, the difference between them was small, but slightly favoured the mixed. Thereafter as home-breds began to come into milk this was reversed, and the gap in favour of the Organic herd increased. And this despite the fact that, due to the policy of using the same bull for both herds, they became more closely related genetically with each successive generation.[1]

Consideration of the figures for bulk yields of crops, summarized in Table 1 on p. 218, in conjunction with the above milk yields, led to the following comments (1962 Report):

'(1) That the value of the extra tonnage obtained by the addition of fertilizers to the Mixed section barely covered the cost of the fertilizers, except in the case of hay.

'(2) That when the crops were used as animal food no benefit resulted from the extra tonnage, rather the reverse. When the findings for the first rotation period, in all departments, were studied as a whole, they led to an evaluation of possible trends, as at that stage, as follows:

TRENDS

'In spite of this greater food intake on the Mixed section, milk production from 1956, when second generation animals were coming into the herds, has been consistently higher from the Organic section herd, whether measured as total milk produced, or production per cow, or production per acre. At the same time the cows on the Organic section have carried better condition, more "bloom",[2] and have shown quite clearly a greater contentment and placidity. This more-milk-for-

[1] See Appendix 10.

[2] It is common for farmer visitors to remark on the superior 'bloom' of the Organic cattle.

There is a point to note here. In the second generation the young beasts had been fed on organic produce since conception. According to Dr. G. Scott-Williamson's interpretation their nurtural inheritance (see 'Science, Synthesis and Sanity', p. 37) had differed from the first original stock, not so fed.

This favourable later performance of the organically fed stock occurred *in spite* of its genetic inheritance via the same bull. The subject here is that a nutritional regime may lead the young to metabolize their food in a different pattern of action.

less-food experience has been one of the experiment's most interesting farm findings to date. It is one which is most marked when milk is being produced from grazing only.[1]

'The next most interesting observation has been the increasingly "self-supporting" nature of the Organic section crops. Those on the other two sections are definitely dependent on the artificial aids they receive. This was to be expected on the Stockless section; but it has been demonstrated, even on the Mixed section that, if fertilizer is omitted from even a small area of any field, the yield of that part of the crop drops well below the yield of the equivalent Organic crop.[2] Conversely, the heaviest yielding fields on the Organic section are those that *have been longest without fertilizers* (thirty-five years in some cases), suggesting that no depletion is taking place. The Organic section crops appear also to be less susceptible to insect pests, and rarely show any deficiency symptoms.[3]

'Other trends during the ten-year period that have been noticeable in the course of ordinary farming operations have included improving "workability" under all weather conditions of the organic fields, and a marked sectional difference, observable only from the third generation, in the proportion of small to large grain in cereals, percentages being of the order of 10 per cent "smalls" on the Stockless crops, 7 per cent on the Mixed, and only 5 per cent on the Organic, as obtained by sieving for seed purposes.

'The comparative health of the livestock of the two sections has naturally been watched with great interest. For the first few generations there was little or nothing to choose between them, both being high, but during the last year or two, indications of loss of stamina have made their appearance in the Mixed stock.'[2]

Following the Summary of his own work,[3] Dr. Milton added:

'In concluding our comments on the period we must come back to one major fact emerging: in both health records and quantity of milk, the Organic herd appears to be superior to the Mixed herd, in spite of the substantially greater quantity of food available to, and consumed by, the latter.

'We looked to the analyses of the fodder from the two sections for an explanation,[4] but there is almost nothing to account for any marked nutritional difference. It is obvious that we must look further

[1] See Graph 5, p. 233.
[2] See p. 240.
[3] See p. 237 and Appendix 6.
[4] For full list of analyses undertaken during the period see Appendix 3.

for the explanation. This is what has prompted us to suggest that the time is now ripe for extending the research into possible unknown nutritional factors.'[1]

The Report concluded with an evaluation of the years of preparation as having helped to answer the question '*what is* happening?', but hardly having begun to answer the question '*why* do these differences (between the sections) exist?' and 'what do they portend?'

Formulating a programme designed to help to find answers to *these* questions was seen as the next step. To this end the following five projects were put forward by Dr. Milton:

(a) An attempt to foreshorten the 'appeal to the cow' (and to the hen and the sheep)[2] by feeding the produce of the three sections to small laboratory animals of quick-breeding turnover for at least ten generations.

(b) The separation of proteins and of amino acids in milk, eggs, and crops, to see whether any differences between the sections can be found which might begin to throw light on nutritional value.

(c) A preliminary search for further accessory food factors.

(d) An initial attempt to find physical or chemical bases for palatability and their significance for nutrition.

(e) Variations in the rumen flora of herbivorous animals which might result from feeding crops of differing quality.[3]

'It is probable,' the Report concludes, 'that other factors in the soil–plant–animal relationship, quantitative or qualitative, which might be associated with nutritional quality, would emerge as a result of pursuing these lines of investigation, for which research material is now [1961] available at Haughley. It is not necessary to start all these projects at once, nor is it proposed to do so. Some we might ourselves be able to follow up through work in Dr. Milton's laboratory, when we can afford some of the new and very expensive equipment. Some could only be attempted in much larger or more specialized research institutions. Each one could be undertaken as a separate research project when the funds and the personnel are available. They would together form an invaluable and integral part of the whole Haughley Experiment, if the hypothesis is accepted, that there is a fundamental soil–plant–animal relationship, and that interference at any point in this cycle will affect the cycle as a whole.'

[1] This the Soil Association, using Haughley research material, was poised to do, awaiting only financial support to further so promising an investigation.

[2] For information on poultry and sheep see next chapter.

[3] For a more detailed elaboration of these proposals see Chapter 7 and Appendices 8 and 9.

4

The Last Eight Years

Unfortunately the experiment finished before the second rotation period could be completed. The reasons for its premature ending were complex and, so far as this report on findings is concerned, no very useful purpose would be served by going into them in detail here. They will be discussed briefly in Chapter 5.

The final eight years have to be divided into two periods, which is also unfortunate. Were it not for this it could have been argued that in providing an eighteen-year continuity of records and analyses this was only slightly less valuable in assessing trends, than the full twenty years which two completed rotation periods would have covered.

In the event, however, a number of changes took place after 1965–6 which broke the continuity in a number of ways. Financial difficulties with escalating labour and research costs were mainly responsible. One effect was to prevent the development of some interesting and potentially fruitful new lines of investigation that were just starting. Worse still, after 1967 all the existing analytical programmes ceased, and during that year the organic closed cycle was partially opened by the application of some imported poultry manure to some of the fields (see p. 243).

Apart from this, the basic farming operations of the three sections continued, and were recorded up to, and including the final report for the year 1969. But even on the practical farming side, the last five years (and particularly the last three) cannot be regarded as a true continuity of the previous fourteen, because of some basic changes in policy that resulted from changes in management and directive (see Chapter 5).

These changes did not all occur at once, but among them were the following:

(a) Suckle-rearing of calves was given up in favour of bucket feeding.

(b) The custom of tripoding hay while green and still in sap, so allowing of slow natural maturing without loss of texture or colour, was given up in favour of drying loose, and bailing direct from the swath. This change resulted in an apparent marked lowering of the quality of the hay produced.[1]

(c) During the winter months the cows were kept in their covered yards night and day instead of going out to pasture for some part of each day, in all weather.

(d) Fertilizer applications to the Mixed and Stockless sections were greatly stepped up, and, in one year (1967), as already mentioned, imported organic treatment was given to part of the Organic section, thus confusing some of the trends.

(e) Much less care was taken properly to compost the FYM and other residues on that section.

(f) The system of grassland management was modified, less attention when strip grazing, being paid to the importance of back-fencing and adequate rest periods. The earlier habit of leaving one ley each winter in foggage (i.e. to stop grazing it in the autumn well before the end of seasonal growth) was abandoned.

(g) Concentrate rations for livestock were compounded and allocated on more conventional lines.

(h) Cultivations, too, became rather more like those of orthodox commercial farming (biological needs giving way to operational speed). In particular, less awareness, than hitherto, was shown for the special needs of the Organic section in the matter of aeration. For example it had previously been demonstrated that spring aeration of organic leys by such devices as pitchpole harrowing stimulated early growth in exactly the same manner as applying a top dressing of nitrogen did to the mixed leys. Similarly, surface cultivation, with aeration of lower levels by chisel ploughing, had proved a better technique for stimulating biological soil activity than deep invertion with a mould board plough.

How much any, or all, of these changes affected functional relationships between soil, plant, and animal it is difficult, probably impossible, to discover, but before presenting any figures relating to average findings for the eighteen-year period (or even comparing the last eight

[1] Under modern farming conditions the technique of curing hay on tripods (or their equivalent, the Scandinavian fence) has been almost universally abandoned except in hillfarms of high rainfall areas. One of the disadvantages of the direct bailing technique which has superceded it, is that even where weathering is reduced to a minimum, the violence of the pre-bailing treatment designed to speed up loss of sap cannot fail to damage cell structure.

years with the first ten), the two four-year periods—1961–4 and 1965–8, need to be considered separately.

PART 1. 1961–1964

Following the publication of *The First Twenty-Five Years* report, some generous donors among Soil Association members provided the gas chromatography apparatus needed to enable Dr. Milton to make a start on projects numbers 2 and 3 of those listed at the end of the last chapter.

In the same year several pieces of electrophoresis apparatus were also obtained, so that by the end of the 1961–2 season a programme for investigating the proteins, amino acids and sugars of crops from the three sections was under way. For the first time, in the analytical work on crops, sectional differences made their appearance (see Chapter 7).

No report on this particular extension of the research is available because the correct interpretation of the differences, and their significance (if any), was still unclear when all laboratory work came to an end.

The previous range of chemical analyses on soils was reduced during the period we are dealing with but that on produce was considerably extended. For example much more work was done on pastures, involving more than 1,000 individual analyses during the growing season, May to September inclusive. The trace element analyses were extended to include straw, and there were more comprehensive cereal measurements. Analyses for vitamin E was added to that for the B group in crops, and vitamin C to the analysis of milk.

In the summary of the analytical work, carried out in 1961 (which Dr. Milton submitted for the report published in 1963), a very interesting pattern of functional relationship emerged between soil and plant, as affecting levels of available nutrients in the soil. This resulted from averaging the analyses for humus, and available nitrogen phosphate and potash, on the stockbearing sections, for the years 1955–60 and presenting the figures for the summer months, April to June in a *rotation* framework. The resulting four graphs, with explanatory comment, are so striking that those for N, P and K are included here, instead of being relegated to the appendix containing the rest of the 1961 analytical summaries.

Available Nutrients

NITRATE NITROGEN

'The highest results are given from May to September, when the soil micro-organisms are most active; this being particularly noticeable on the organic arable fields (less so under ley).

'Graph 2 shows the rotational trend if we average out the figures for the months April to June each year. The uprising when the fields are under pulses is much in evidence, with a fall-away on the return to cereal crop.

GRAPH 2

AVAILABLE NITRATE NITROGEN 1955-1960, average three months April, May and June

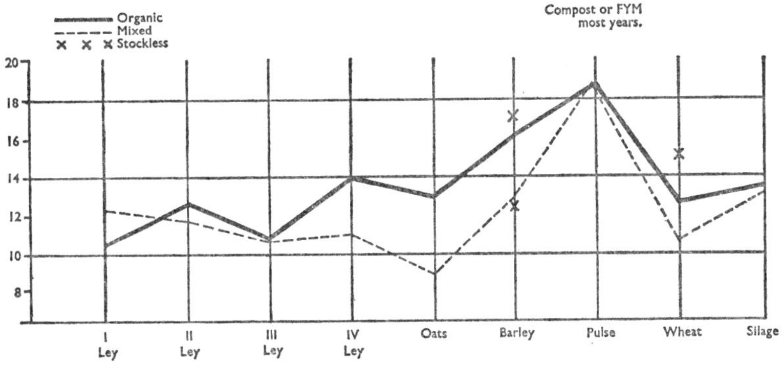

PHOSPHATE

'The tendency for highest levels in the spring and summer months is quite apparent, although on the fertilizer sections, the picture is complicated by the additional effect of inorganic phosphate addition.

'Some of the rotational trends in phosphate levels are most interesting. If we relate phosphate levels to the crop, it is seen that the average levels on mixed section are only slightly higher than on the organic, despite the annual additions of phosphate fertilizer on the former.

'Variations in average P_2O_5 level for the three growing months when plotted against crop show an inverse type relationship of graph (Graph 3).

GRAPH 3

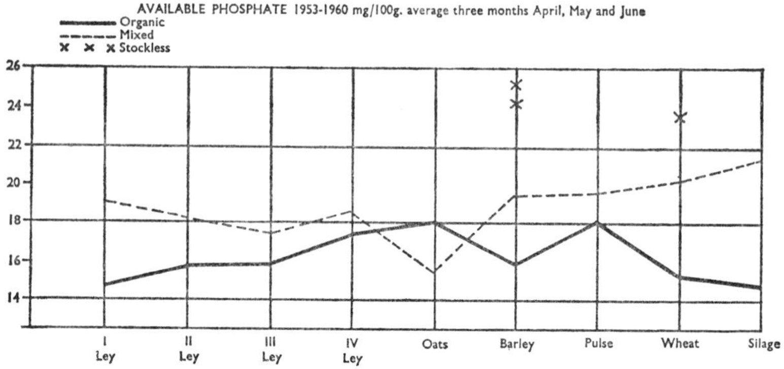

AVAILABLE POTASH

'The yearly figures show highest potash in June on all sections; the rotational trend is definitely connected with crop as is amply seen in Graph 4, the level being low under the first two years of the ley, and then rising. Two corn crops rather lower the level, which nevertheless makes a high peak with pulses, and this falls again when the soil is put back under ley.'

GRAPH 4

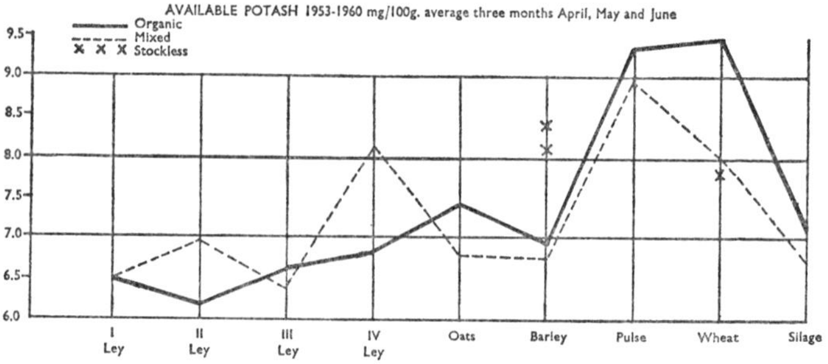

Dr. Milton's view was that because leys consisted of plants in continual growth (albeit slight during the colder weather) there was not the same dramatic 'stop-go' alteration in functional pace in all-growth processes, e.g. plant and soil, as is the case—perhaps *has* to be the case—with a short-term arable crop such as a cereal.

In the second half of 1963 some small, short-term, pilot nutritional experiments using Haughley produce were attempted in collaboration with the Huntingdon Nutritional Research Laboratories. The outcome with small-animal feeding tests was very inconclusive, but one experiment, relating to the *utilization* of food by cattle, produced interesting pointers.[1] This work, however, was in too early a stage to be given more than a passing reference in Dr. Milton's summary of the 1962–3 Analyses (which formed part of the annual report, published in 1964). This summary concluded with the following paragraphs:

'Perhaps some of the most interesting extensions to our work have been the investigations into the composition of the sugars in the crops, and we hope to be able to extend this work in future to a quantitative basis.

'Whereas in the past we have been concerned mainly with the crude nitrogen content of crops, we have attempted to differentiate the protein fractions, and in particular to determine the non-protein fraction of the cereals. [For total protein, see Appendices 5 and 6.]

'We have extended our work on muck and compost analysis to include trace element contents so as to complete the story on all the substances concerned in the cycle.

'As 1962 was the first year of a more comprehensive attack in order to obtain leads concerning quality in foodstuffs, it is a little premature to give opinions as to results, particularly in view of the fact that there were a number of crop failures during that year.

'Our future work will be concerned with following up more refined analyses of crops and pastures. Considerable headway has already been made with methods for determining the amino acid differentiation in the products.

'In 1962 we made a start on the gas chromatographic separation of the fatty acids in the crops. Some forty-five fatty acids have been followed up on each of the crops analysed on each of the sections, and it is proposed eventually to make a statistical analysis and interpretation of the results obtained. There are some very striking differences in the percentage content of the various fatty acids in the products from the various sections.

'We have had some confirmation of this from the work now being carried out on the 1963 crops, and on the material from the fistulated heifers,[2] but it is still too early to comment.

'It is of great interest to be able to compare results for the two years over such a wide range.

'It certainly cannot be said that any deficiencies are apparent in the

[1] See Appendix 9, pp. 355–6 for short report of this work.
[2] See Appendix 9.

TABLE 2

DATA	1962				1963			
	Organic	Mixed	Difference: Organic higher	Difference: Mixed higher	Organic	Mixed	Difference: Organic higher	Difference: Mixed higher
Total cow-days	4,790	5,213		423	5,237	5,636		399
Total concentrates fed (lb.)	8,521	13,158		4,637	11,506	16,183		4,677
Total concentrates per cow-day (oz.)	28·5	40·4		11·9	35·1	42·3		7·2
Milking cow-days	3,940	4,067		127	3,762	3,877		115
Concentrates fed to milkers (lb.)	6,753	10,144		3,391	8,068	12,383		4,315
Concentrates per milker cow-day (oz.)	27·4	39·8		12·4	34·6	51·1		16·5
Total milk (lb.)	73,899	69,514	4,385		85,205	78,288	4,176	
Milk per milker cow-day (lb.)	18·5	17·1	1·4		22·6	20·2	2·4	
Milk per acre (lb.)	939·8	916·2	23·6		1,083·7	1,031·2	52·5	
Milk per lb. concentrates fed to milkers (lb.)	11·0	6·9	4·1		10·5	6·3	4·2	
Milk per lb. total concentrates fed (lb.)	8·7	5·3	3·4		7·4	4·8	2·6	

produce from the Organic section as a result of not applying chemical nutrients in artificial form.

'It would not be proper to finish these observations without commenting on the fact that once again the milk yield from the Organic cows was considerable higher than that from the Mixed herd, and reference to the crop yields shows that there was considerably less feed produced on the Organic section.

'It becomes increasingly obvious, therefore, that we must investigate the problem very fully from all aspects in order to clarify the reasons for this consistent difference and to show whether it is in any way related to the composition of the foodstuffs.'

Inevitably the reports published in 1963 and 1964, concerning farm records and observations, largely centred on this puzzle—analysis of the relationship between food intake and milk production in 1961 showed that although the number of cow-days, the concentrates given per cow, and the concentrates given per gallon, were all higher on the Mixed section, the total milk, the average per cow, and the amount produced per lb. of concentrates fed, were all greater on the Organic section.

From 1962, total milk production *per acre* was added to the above data. The results for 1962 and 1963 are shown on p. 230.

The table on p. 232 gives the 1964 figures, and the mean of the previous three years.

The fact that even the milk per acre was greater on the Organic section could indicate that the summer pasture food intake may have been a bigger factor than the winter concentrates. Certainly, in most years, the gap in production between the two herds was greatest during the grazing season, in spite of the obviously greater quantity of herbage and lusher growth on the 'M' leys.

The 1961 season illustrated this in a more than usually dramatic way (see Graph 5, p. 233).

Note the jump between the third and fifth four-week period when the cows went out to grass. The first-year leys in question, during April, May and June, produced a weekly average of 14 gallons per cow on the Mixed section, and 20 gallons per cow on the Organic.

The field observations recorded on the 1961 leys (published in the 1963 report) stated:

'The mild February and March produced an excellent early bite on the Organic leys—the cows were sleeping out and strip grazing in the first week in April. The Mixed leys on the other hand were still at a standstill in March and did not begin to "move" until they were

TABLE 3

DATA	*1964*				*Mean of previous three years*			
			Difference				*Difference*	
	Organic	*Mixed*	*Organic higher*	*Mixed higher*	*Organic*	*Mixed*	*Organic higher*	*Mixed higher*
Total cow-days	4,860	5,479		619	5,274	5,620		346
Total concentrates fed (lb.)	10,414	14,384		3,970	13,628	18,653		5,025
Total concentrates per cow-day (oz.)	32·0	40·8		8·8	41·4	53·1		11·7
Milking cow-days	3,677	4,022		345	3,689	4,020		331
Concentrates fed to milkers (lb.)	7,646	11,034		3,388	7,855	14,005		6,150
Concentrates per milker cow-day (oz.)	33·7	43·9		10·2	34·1	55·7		21·6
Total milk (lb.)	78,236	73,784	4,452		80,966	76,452	4,514	
Milk per milking day (lb.)	21·3	18·4	2·9		21·1	19·0	2·1	
Milk per acre (lb.)	995·0	972·2	22·8		1,029·7	1,007·6	22·1	
Milk per lb. concentrates fed to milkers (lb.)	10·3	6·7	3·6		10·3	5·5	4·8	
Milk per lb. total concentrates fed (lb.)	7·6	5·1	2·5		5·9	4·1	1·8	

top dressed, then they caught up to the Organic leys. This behaviour pattern of the leys does, however, appear to depend on favourable temperature and moisture conditions for biological development. For instance in the cold dry spring of 1962, the fertilized Mixed leys started earlier than the Organic ones.'

GRAPH 5

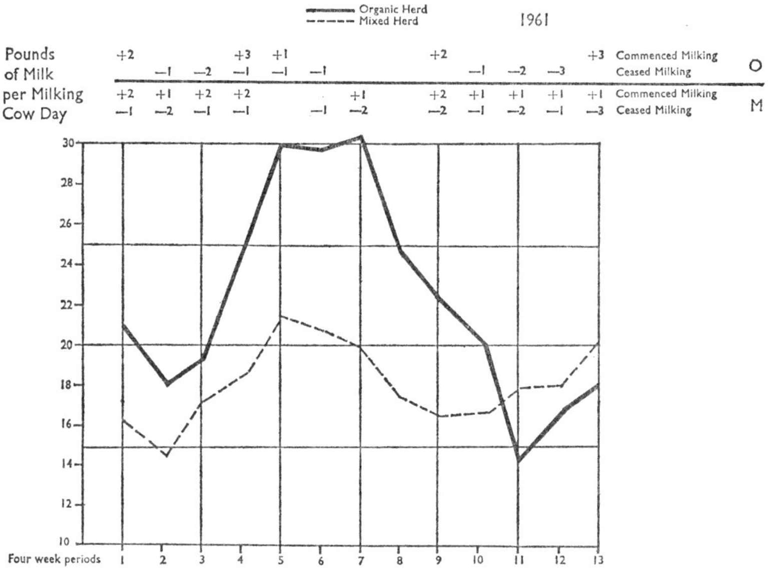

Observations on the behaviour of the leys in 1963 (see report published in 1964) confirmed and amplified the 1962 findings and were reported as follows:

'The very late spring caused a slow start to the "O" first-year ley, and the ' O" herd were a week behind the other going out to grass, but when the growing weather came at last everything grew so fast that the cows could not keep up with the growth and a section had to be cut to stop it being spoiled. The same thing occurred on the "M" first-year ley, except that the surplus growth had to be cut a few weeks earlier.

'An interesting observation occurred in grazing behaviour on this "M" ley. When the spring compound fertilizer was applied a strip across the centre of the field was deliberately omitted. When the cows were first turned out in this field they immediately found this strip and grazed it bare before feeding elsewhere. This difference in palatability of the treated and untreated herbage disappeared later in the season.'

Dr. Milton's report mentions the extension of analytical work carried out on the pastures. The superiority in total solids content of the Organic herbal leys helps to explain the phenomenon of more milk for less bulk. Samples taken on 2 May 1963 for example, showed:

Section	*Dry matter* %
1st year 'M' ley fertilizer treated area	11·8
1st year 'M' ley, no fertilizer	13·0
1st year 'O' ley	18·2

A week later both areas on the 'M' ley yielded 13 per cent dry matter and the 'O' ley 26 per cent. That is to say that the 'M' cows sometimes have to eat twice as much in order to get the same quantity of food.

This factor, however, was not constant, and did not apply to grain, pulse or hay (though silage often followed the pattern of the leys) so that by itself it was insufficient to explain the production differences between the two herds.

Correspondence following publication of the first report, suggested that the most likely cause lay in genetic differences. This theory had also, of course, occurred to the management and directors. Therefore a large part of the report published in 1963 was devoted to a very detailed herd history. Breeding and performance of every animal in both herds, including the foundation heifers, was set out in various different ways designed to reveal a genetical causation of the findings, if such existed.

The theory failed to stand up to this examination.[1]

Two other possible factors considered were (a) the use of dried seaweed as a trace mineral supplement for the Organic herd, against normal mineral licks for the mixed,[2] and (b) the wide range of deep-rooting grasses, legumes and herbs used for the Organic leys, against the simpler more conventional mixture on the Mixed.[3]

The only way to investigate the effect of these two possibilities was

[1] The full details are too long for this digest. The salient points are given in Appendix 10. Readers wishing to study the available data on this aspect in greater depth are referred to the annual report published in 1963 available on loan on application to the Soil Association. (See Appendix 12.)

[2] For analyses see Appendix 2, p. 308.

[3] For composition see Appendix 2, p. 300.

to eliminate, in turn, each of these two differences. This it was intended to do, had the original experiment continued long enough.

In the period being dealt with (1961–4) the general impression persisted (see Chapter 3) that the Mixed herd was dropping behind the Organic in the quality of stamina. Since one of the symptoms of this quality is considered to show up in breeding records, a review was made in 1964 of all cattle sold from both herds, from the start in 1948 up to the (then) current year. The resulting analysis shows that of the forty-one cows and heifers sold during this period from the 'O' herd, seven were culled for failure to breed (17 per cent). Of the forty-six cows and heifers sold during the same period from the 'M' herd, sixteen were culled for failure to breed (34·8 per cent). These figures appear to bear out the general impression. The 'O' cows also averaged a longer working life. On the other hand, 'in the period 1961–4, of the sixty-eight births in the "O" herd ten were still-born as against five in the seventy births of the "M" herd.'[1]

Poultry

The one place during this period where the use of fertilizers on the Mixed section succeeded in increasing stock-carrying capacity was in the case of poultry. The extra bulk yield of grain enabled fifty more hens to be kept in the Mixed flock than in the Organic. *Total* egg production has thus been higher on the Mixed. Numbers of eggs per bird was approximately the same in both flocks, small differences occurring in both directions. But since, as with the cows, the Organic flock received less food than the Mixed, the relationship between food intake and egg production usually followed a similar pattern as in the case of milk.

The foundation chicks (Light Sussex) were bred by the late Mr. Golden of Leire in Leicestershire, whose flocks are renowned for their stamina, and whose methods (summarized in what he called the 'Four Gs' of grass, grain, greens and grit) were followed as closely as possible with the Haughley poultry.

Both flocks were kept on free range the whole year round (usually on the third year of the four-year leys).

As with cattle, they were fed almost exclusively on home-grown produce. The evening feed always being of whole grain. In summer, three morning feeds a week were also given as grain, the other four as mash. In winter, six morning feeds a week were given as mash.

[1] J. F. Ward, *The Haughley Experiment 1952–1965*, see Chapter 5.

The average degree of production range is sufficiently indicated by the following table:

	1961 O	*1961* M	*1962* O	*1962* M	*1963* O	*1963* M	*1964* O	*1964* M
Average number of birds	216	270	194	242	229	273	230	293
Average number of eggs per bird	172	151	159	159	152	155	126	134

The average annual egg production per bird over an eleven-year period was 'O' 163; 'M' 160.

The figures for the number of lb. of food consumed for each dozen eggs produced were:

	1962	*1963*	*1964*
Organic	7·7	7·4	8·9
Mixed	8·3	8·4	9·3

The health and breeding records appeared to be rather contradictory. In most seasons a higher proportion of hatching eggs were fertile in the Mixed flock. Numbers of live chicks, and chick mortality were roughly the same in both, but adult mortality was significantly lower on the Organic flock.

This might be accounted for by the method of open-air housing in all weather ensuring the survival of the fittest, i.e. those with the best stamina. One of the reasons for the adoption of this system by Mr. Golden was deliberately to weed out the unfit. Certainly the two commonest causes of adult mortality in the Mixed flock were respiratory troubles, and heart failure (often from overweight despite free ranging).

It might be worth while at this point to recall that more than one customer of Haughley pullets and older hens at the annual cull have reported that those from the Organic section have a much higher survival rate and give more eggs when submitted to more intensive conditions, than do those from the Mixed section. On one occasion a customer bought approximately fifty hens from the Organic section and fifty from the Mixed and put them together. Then it occurred to him to have an 'expert' pick out unproductive hens. He picked out about twenty-five—every one was from the Mixed section. Another customer once also had some birds from both flocks, and put them with a third batch from a commercial hatchery. Disease was introduced with the latter. He eventually lost both the commercial pullets and also those from the Haughley mixed flock, but all those from the Organic flock survived.

Sheep

There is little to report about sheep for the period, since the foundation ewes for the two tiny Clun Forest flocks were not bought until 1962, so 1963 was the first year any home-bred lambs were mated.

The breeding and production record in 1964 was as follows:

	Organic	*Mixed*
Total ewes and hoggets mated	17	16
Number of lambs reared	25	22
Wool clip in lb.	108	96

Crops

The bulk yields of crops during the period followed very much the same levels, range of variations, and general pattern of the first rotation period given in Chapter 3. So it has not been considered necessary, to include them here for the years 1961–3. The picture can equally well be given by the following table showing the yields for 1964 and the mean of all the previous years. The figures in the second column represent not only the number of years averaged, but also, in the case of cereals and pulses, the number of generations of home-grown seed.

A few additional field observations, taken from the annual reports covering the period, are of interest, and are quoted here:

'There is some indication that the dependence on "aids" in the "M" and "S" sections [referred to in Chapter 3, p. 222] applies also to weed infestation, and to insect pests, though caution is obviously required here. Nevertheless, weeds do appear to be increasing on those two sections where herbicides are regularly used, and it would be difficult now to keep them in reasonable control in the "M" and "S" cereals without resorting to sprays. The "O" fields present much less of a problem in this respect.

'Concerning pests, the most striking observations have occurred in the case of weevil, to which, both in 1960 and 1961, the Organic crops appeared much more resistant. In 1960 the newly sown Mixed lucerne would have been a complete failure if it had not been sprayed against weevil, and in 1961, only spraying saved the Mixed pea crop from destruction by this pest, though it should be stated that the Mixed peas followed an old lucerne ley, and this may well have resulted in an

TABLE 4

CROP	No. of yrs.	ORGANIC		MIXED		STOCKLESS		Cost of fertilizer per acre 1964		Cost of spray per acre 1964	
		Mean cwt./acre	1964 cwt./acre	Mean cwt./acre	1964 cwt./acre	Mean cwt./acre	1964 cwt./acre	M	S	M	S
								s d	s d	s d	s d
Wheat (Atle)	12	22·4	25·2	26·7	25·9	25·4	29·9	92 2	93 10	—	—
Winter Oats (S.147)	11	22·2	30·6	24·5	38·3	—	—	91 3	—	27 6	—
Spring Oats	3	16·9	25·2	24·4	24·5	—	—	56 4	—	27 6	—
Barley (Rika)	9	27·3	19·2	30·1	29·7	31·6	28·2	71 2	63 0	—	15 11
Barley 2nd crop	9	NG	NG	NG	NG	24·5	23·5	—	89 7	—	27 6
Beans	12	11·7	13·6	13·8	16·5	16·8	14·0	79 4	67 5	—	—
Peas	12	9·4	12·3	6·8	15·3	NG	NG	79 4	—	—	—
Maize	2	223·7	120·8	399·2	82·0	NG	NG	150 9	—	—	—
Kale	2	116·1	83·0	537·6	184·0	NG	NG	135 7	—	—	—
Lucerne Silage	3	102·7	NG	154·0	NG	NG	NG	—	—	—	—
Lucerne Hay	3	26·4	NM	39·9	17·5	NG	NG	47 8	—	—	—
Ley Silage	2	85·6	NG	277·0	245·2	NG	NG	64 4	—	—	—
Ley Hay	3	30·9	30·1	50·4	21·5	NG	NG	—	—	—	—
Arable Silage	3	85·5	137·0	156·3	187·8	NG	NG	40 2	—	—	—
Sugar Beet	3	NG	NG	NG	NG	252·6	248·7	—	178 0	—	—
Red Clover Seed		NG	NG	NG	NG	1·5	1·7	—	—	—	—

Notes: Mean is to 1963 inclusive.
Organic Section receives no fertilizer or sprays.
NG = Not Grown.
NM = Not Made.

extra heavy infestation on that section. The fact remains, however, that neither lucerne nor peas were attacked on the Organic section.

'A further indication that this was not due just to chance, was that in the oat-and-pea mixture for silage in 1961, the "M" crop was not sprayed and practically every pea was destroyed by weevil. Little if any damage took place on the "O" crop.'

Note: In 1962, the trouble was not weevil but pigeons, who destroyed the entire pea crop on the Organic section and all but 5 cwt. on the Mixed section.

This necessitated purchasing new seed-peas for the 'O' section for most of the 1963 sowing. Enough remained of the 1961 crop, however, to seed a small area with this, in order to ensure that no break would take place in the continuity of seed generations. Thus the 1963 crop was grown from approximately 2 acres sown with newly bought seed, and an acre with two-year-old twelfth-generation home-grown seed. The superior yield of both peas and haulm from the latter was very striking, and also unexpected. Both were, of course, the same variety and grown in the same field.

Yield per acre of the 1963 Organic pea crop from area sown with:

	Bought seed *cwt.*	*Home-grown seed (twelfth generation)* *cwt.*
Peas	10·8	16·6
Haulm	19·0	33·0

Finally, the observation that the soil of the Mixed fields had become dependent on artificial aids (mentioned in Chapter 3) continued to be noticed in the following years. In 1963 and 1964 an attempt was made to test this impression on the crops of kale, maize and turnips. The result is shown in the following table.

It will be noticed that there was a very considerable seasonal difference in the kale yield on both sections, but in both years the weight of all three crops on the Mixed section was much heavier than the Organic ones, *except where fertilizer was intentionally omitted.* Here, in spite of receiving farmyard manure the yield was *less* than that of the Organic crops though the latter were growing in fields that had had no fertilizer for fifteen to twenty years, but only compost made from the section's own wastes.

The estimates of yields were made from a large number of measured samples taken, and weighed by Mr. Ward,[1] Assistant Scientific Officer, and the results converted into tons per acre.

[1] See Appendix 1.

TABLE 5

Crop and year	*Organic*	*Mixed*	
		With fertilizer	*Without fertilizer*
	tons/acre	*tons/acre*	*tons/acre*
Kale			
1963	14·6	28·6	10·8
1964	4·1	9·2	3·1
Maize			
1963	9·7	11·7	4·7
1964	9·9	4·1	3·4
Turnips			
1964	8·3	9·8	6·3

PART 2. 1965–1968

From 1965, under the new management (see Chapter 5) the data in the annual reports, while still following approximately the same pattern of presentation as previously, nevertheless differed in two major respects. First, field observations were no longer accompanied by analytical findings, with the result that interpretation of the former, in those cases where analyses would have provided it, was lacking. This is not to say that all analytical work ended in 1965. Dr. Milton's investigations of the research material (a wealth of which was by that time available, some of it unique), although progressively curtailed, did not end altogether until 1967, but the findings were no longer linked, as in the previous manner, to field observations.

Second, the record system inaugurated by Douglas Campbell differed from the one in current use till that time, quantitative production being given more emphasis than behaviour patterns. These two changes possibly make it undesirable to compare findings between the two periods, i.e. before and after 1965. Certainly any such comparisons should be made with caution. Nevertheless, in making my digest of the annual reports for the later period, I have summarized them in such a way as to bring out items relevant to earlier findings, whether apparently confirmatory or contradictory. Thus the selected quotations from the four annual reports which follow (1965–8), include examples of both these categories, and have been arranged under headings that should enable some measure of cautious comparison to be made with those of earlier years. A few comments and/or explanations have been added where they seem called for, and are printed in italics, to distinguish them from narrative passages.

Before proceeding to examine these quotations in any detail, it should be noted that although the experiment did not officially finish until early in 1970, the report for the last year (1969) was not available when the decision to alter the whole basis of the original research was taken *and* implemented. The field observations and production figures for 1969 are, therefore, given in Part 3 of this chapter as an addendum.

Field Observations (arable)

1965

'The Organic wheat was thin but stood up well to the weather, whilst both Mixed and Stockless were badly laid. The yield of the latter, at one time a most promising crop, was thereby greatly reduced.

'Organic spring oats withstood the weather well but much of the Mixed spring oats was laid. Yields of winter oats were below average, but those of spring oats were above.

'The first sowing of kale on the Organic section was destroyed by pigeons and flea beetle and was resown six weeks later. Although an excellent take was obtained the crop never caught up with the Mixed kale. Although the yield of the latter was estimated to be almost double that of the former, its feeding value was much reduced because of its high proportion of stems, 54·6 per cent, as against 31·1 per cent.

'The yield of the Organic silage was, as usual, less than that of the Mixed, but yields of both sections were well above the farm average. Similarly, the yield of Organic hay was lower than the Mixed, but both yields were above average.'

1966

'All [wheat] yields were well above average for the farm, being approximately 37 cwt. per acre for the Mixed and Stockless and some 9 cwt. less for the Organic. The incidence of bunt has increased considerably on the Organic. In 1965 9·94 per cent of the ears sampled (186) were bunted and in 1966 this had risen to 22·2 per cent. In neither year was it possible to find a single bunted ear on the other sections.'

(*See p.* 219 *for the history of the bunt infection in the Organic wheat. Bearing this in mind, the steady level of the yields of Organic wheat, even unto the twentieth generation, are noteworthy.*)

'Spring oats were sown on 9–11 March and came away well. The Organic oats on Lane field had to compete with a very heavy investation of charlock. The crop stood up well to the weather. The Mixed oats promised a very heavy crop but were badly laid following the rain of 17 June. The Organic oats yielded 26·5 cwt. per acre and the Mixed 35·7 cwt. Both yielded better than in previous years.

'Barley crops as a whole were heavier than in 1965, which in itself was an above-average year. The Organic barley which produced 31·4 cwt. per acre appeared to have had too much nitrogen from the sheep-folding previously. As a result it was partially laid and ripened most unevenly. The Mixed barley, which yielded equal to that of the Organic, reflected the poor drainage of the field, Big North, and was partially laid. The Stockless barley gave the heaviest yield particularly the second-year barley on Further Break, where it was fairly extensively laid. The undersown red clover stood out strikingly at harvest time on a strip of barley from which the fertilizer dressing had been withheld.

'The Mixed kale was very robust and much taller than the Organic. Its yield was estimated to be some 30 per cent heavier, although it had a slightly larger percentage of stem.

'Turnips grew away well and the Mixed were noticeably more leafy. They were attacked in July and August by mildew. At the time of sampling for estimating the yield some 6 per cent of the Mixed turnips had rotted. There was practically no difference in total yield, although the Mixed had some 50 per cent heavier top growth.

'The yield of the Mixed mangolds was more than 50 per cent greater than that of the Organic. This difference may have been accentuated by the fact that the Organic mangolds during September developed yellowed and bronzed outer leaves with a certain amount of necrosis. This was very severe in several places across the field. Only very slight symptoms were noted on the Mixed mangolds in the adjacent field.'

(*It is regrettable that the curtailment of the analytical work prevented investigation into the cause of this observation.*)

'In recent years a customer for mangolds for wine making insists, after experiment, in buying only from the Organic section, on the grounds that they give a "better" fermentation.'

(*No explanation was sought for this observation either.*)

1967

'On both sections the turnips for sheep feed were satisfactory and up to average crops were obtained. Once again the root crops on the "O" section showed discolouration of the leaves, oranging of mangolds and purpling of turnips which may be due to a trace element deficiency.'

(*In earlier years this speculation could not have arisen—Dr. Milton's corollary work would have determined it one way or the other.*)

'Kale on the "O" section was short and pale in colour compared with the "M" and yet the cattle ate all of it whilst rejecting the tougher stems of the "M" section. Feeding commenced on 10 October and lasted until 16 December.'

(*This whole question of palatability and animal-consumer preference needs, and (had the experiment continued) would have received, much more research. See also p. 252 below.*)

Although 1969 has been taken as the final year of the Haughley Experiment it is more than questionable whether it should not have been officially recognized as having come to an end in 1967, for in that year the Organic closed cycle was, for the first time, thrown wide open by the application to almost every crop on the section of imported poultry manure pellets and/or maxicrop (liquified seaweed). The only commentary concerning this, in the 1967 report is the following apologia:

'The use of an organic foliar spray and poultry-manure pellets on the "O" section was decided upon during a period of considerable uncertainty concerning the future of the Soil Association and Haughley Research Farms Ltd. It is not intended that any further use will be made of such materials. Prior to the use of poultry manure it had been tested for the presence of antibiotics. The result was negative.'

(*It is not explained who 'decided' to make this change in the first place, nor who does not 'intend' to repeat it. Both 1968 and 1969 did, in fact, revert to the original rules and pattern, except for seed dressing (see below), which is why it was decided to include them in this account, but the 1967 episode indicates the beginning of the controversy within the management which contributed to the termination of the experiment in 1970. See next chapter.*)

1968 (Field observations continued)

'The wheat germinated and grew well but took a beating on the "M" and "S" sections from rain and wind in early August. The seed on the

"O" section was dressed with "Harvesan" at the rate of 4½ oz. per cwt. in order to counteract covered smut which had been troublesome from the previous year despite treatment with formaldehyde.'

(*'Harvesan' is a mercury dressing, so this was the first time, since the start of the experiment, that a toxic and cumulative chemical was applied to the Organic section. Covered smut is another name for bunt.*)

'Oats were drilled between 2 and 5 March. The take was good on both sections but the "O" section did not maintain the early promise of growth and looked like being a very poor crop until late July when it started moving again. The "M" crop was badly laid by heavy rains in mid June.

'Rika barley drilled in early March, the barley germinated well on all sections. The "O" section looked very poor after germination but pulled together in June.[1] Lodging of barley on the "M" and "S" section took place in June and July. The "O" barley did not go down.

'The bean crop on both the "M" and "S" suffered from "chocolate spot" whilst the "O" section had a much less severe attack. The "O" beans gave a much larger and better-quality straw than the other two sections.

'The "O" section kale and mangolds were particularly poor yielding. Part of the trouble was caused by ineffective drainage in Wood field (west end).'

(*This observation on the 1968 kale has relevance to Chapter 5.*)

'Lucerne on the "M" section now into its fourth year is beginning to show signs of deterioration due to weed and grass growth. An attack of eelworm was very noticeable in July but appears to have been successfully controlled by the application of salt in one area and FYM in another.'

(*The fact that a localized application of farmyard manure successfully controlled an eelworm attack is an important observation, since it can only have an ecological explanation, and was almost certainly brought about by predacious fungi.*)

Leys

Only two field observations of note, concerning leys, occur in the four reports—the first, in 1965, states:

'Following on the dry summer and autumn of 1964 the springing of the leys was slow, particularly on the Organic fields where it

[1] This pattern of growth (root development being produced before top growth) was always typical of the Organic cereals. (See Chapter 6, p. 267.)

became necessary to bring in a load of organically grown mangolds to supplement feed. As the season advanced the leys grew faster than the stock could consume the grass, most markedly on the Mixed section. As a result, a portion of Big South and all of Sawpit (both Mixed fields) and a portion of the Organic first-year ley were cut for hay. In this respect it was repeatedly noted, when taking pasture samples, that the first- and second-year leys on the Mixed section had a considerable amount of dead material left unconsumed from earlier grazing periods. But on the comparable Organic leys the cattle kept on top of the grass and did not allow this to happen.'

(*Palatability seems again to be involved here, for even when cattle were given no alternative food, they refused to graze the Mixed leys close.*)

The second observation occurs in the 1968 report, and is of special interest since it confirms much earlier findings, and underlines the need for further research.

'It is again apparent from the increase in milk yields which occur after the two herds are put out to grass in the spring that there is more to grazing than a difference in "green bulk". The "M" grazing was about 5–6 inches tall and a dark, lush blue-green in appearance, the "O" side was 2–4 inches high, sparse and a lighter green. The percentage increase in milk yield, after the first week's grazing, taking the previous week's figure as a base line, was 30 per cent for "O" and 20 per cent for "M". There are many factors which could be partly responsible for these figures but there is still an intangible factor for "quality" in the "O" grazing which requires much more detailed study.'

Milk Yields

'Annual report Table No. IX gives the intake of concentrated feed and the output of milk in respect of the Organic and the Mixed herds, and compares them with the mean of the previous four years. The previous trends have for the first time been reversed. The total production of milk is approximately 700 gallons greater from the Mixed herd and production of milk per acre is in favour of the Mixed herd by 125·5 lb. However, milk produced per lb. concentrate fed to milking cows remains in favour of the Organic herd by 3·4 lb.'

(*It could have been too early to deduce from this a 'reversal of trend'. The milk production per acre in the following years was twice higher on 'M', and twice (if 1969 be included) higher on 'O'. The actual figures being:*)

lb. production per acre	*1966*	*1967*	*1968*	*1969*
Amount by which 'O' is greater	—	279·1	—	182·2
Amount by which 'M' is greater	83·5	—	390·5	—

The following tables, taken from the four annual reports, 1965–8, show total production, and production per cow:

TABLE 6

Year	*Herd*	*Total Milk lb.*	*Average number of cows in herd*	*Average milk per cow*	*Average milk per cow, for year*
1965	Organic	70,682	12·6	6,271 (10 years)	5,609
	Mixed	77,720	13·6	5,708 (10 years)	5,715
1966	Organic	79,709	12·6	6,098 (11 years)	7,762
	Mixed	78,091	13·4	5,599 (11 years)	6,529
1967	Organic	84,918	13·5	6,121 (12 years)	6,290
	Mixed	60,856	11·6	5,618 (12 years)	5,246
1968	Organic	78,077	13·6	6,135 (13 years)	5,741
	Mixed	98,601	16·9	5,590 (13 years)	5,834
1969	Organic	94,550	14·5	6,162 (14 years)	6,520
	Mixed	81,285	15·1	5,575 (14 years)	5,383

Throughout this period the milk produced per lb. of concentrates fed continued to be greatest for the 'O' herd with the one exception of 1966 when both herds were the same.

(*The total concentrates fed during the four years are difficult to link, either with total herd yield or with numbers of animals. In 1965 the 'M' cows received 3,528 lb. more than 'O'. In 1966 'O' received 763 lb. more than 'M'. In 1967 'O' again received more than 'M' by 3,784 lb. In 1968 'M' once again received more than 'O' by 5,433 lb.*

These wide differences seem to need more explanation than can be deduced from the available figures. Without this, it is not possible to suggest how comparable they are with the earlier period, but we give on p. 247 the 1968 table in full, to match that given for 1964 in Part 1 (p. 232).)

TABLE 7

Milk production table from the 1968 Report	MEAN 1961–7				PRODUCTION 1968			
			Difference				*Difference*	
	Organic	*Mixed*	*Organic higher*	*Mixed higher*	*Organic*	*Mixed*	*Organic higher*	*Mixed higher*
Total cow-days	4,856	5,150	—	294	5,078	6,387	—	1,309
Total concentrates (lb.)	13,064	15,338	—	2,274	15,561	20,994	—	5,433
Total concentrates per cow-day (oz.)	43·0	47·4	—	4·4	48·6	52·6	—	4·0
Milking cow-days	3,651	3,963	—	314	3,691	5,364	—	1,673
Concentrates fed to milkers (lb.)	9,472	12,127	—	2,655	13,390	18,712	—	5,322
Concentrates per milking cow-day (oz.)	41·4	48·1	—	6·7	58·0	55·7	2·3	—
Total milk (lb.)	78,712	72,067	6,645	—	78,290	97,580	—	19,290
Milk per milking cow-day (lb.)	21·5	18·5	3·0	—	21·2	18·2	3·0	—
Milk per acre (lb.)	1,001·2	968·6	32·6	—	995·6	1,286·1	—	290·5
Milk per lb. concentrates fed to milkers (lb.)	8·8	6·1	2·7	—	5·8	5·2	0·6	—
Milk per lb. total concentrates fed (lb.)	6·3	4·8	1·5	—	5·0	4·7	0·3	—

For the 1969 production figures see next chapter.

Health. The 1966 report contained the following:

'The Organic herd was singularly free from accident and disease in 1966, no outstanding occurrences being recorded.

'On the Mixed section, one heifer was barren from the first bulling, having five services including hormone injections to no avail, finally being sold as barren. The incidence of milk fever was greatly reduced, only two cases being treated. One cow (Beauty VII) responded to two-bottle calcium injection treatment, while the other (Chloe IV), in spite of a prolonged series of injections, collapsed and died on 7 May 1966.

'Only one serious case of mastitis occurred, responding readily to antibiotic treatment and iodine wash.'

(*In 1968 a case of milk fever is recorded in the Organic herd, and in the Mixed herd two cows were lost from Johne's disease. This disease was introduced by a young bull bought in the very early days. It cropped up from time to time thereafter, but there were never any losses from it on the 'O' herd, which is not without interest, seeing how very contagious this complaint is. No inoculations were given to either herd.*)

Egg Production for the period was as follows:

Year	*Average number of eggs per bird*		*Pound of food per dozen eggs*	
	O	*M*	*O*	*M*
1965	170	153	6·70	7·70
1966	151	151	7·68	7·70
1967	158	160	7·07	7·29
1968	178	173	6·12	6·30
Average for ten years	162	160	6·80	7·90

The *1968* report comments on the production for that year:

'This has been a particularly good year for eggs. In both sections the management was improved and as a result less food was used and a greater total production per bird effected. Most of the eggs were sold to the wholefood shop in London.'

Wool Production

The *1965* report shows that the wool clip for that year was 5 lb. less from the 'O' flock than from the 'M' from the same number of sheep (twenty-one in each flock) and was inferior in quality compared with 'M', as judged by an expert of the wool marketing board.

The *1966* clip was again slightly heavier on the 'M' flock—5.30 lb. per sheep against 5·20 lb. in 'O'. Quality was also again better from 'M' as reflected in the price received—55·27d per lb. 'M', and 55·06d per lb. 'O'.

In *1967* almost the same difference in quantity, as in 1966, is recorded. No mention was made of quality.

In *1968* the quantities were reversed, the 'O' flock yielding 6·9 lb. per sheep, and the 'M', 5 lb. Again no mention was made of quality.

Total livestock populations for each of the last eight years are given in Appendix 5.

Crop Yields

The table on page 250 gives the 1968 yields, together with the mean for previous years, and corresponds to the similar table for 1964, plus means, given on page 237.

The explanation given in the latter case, concerning the figures in column 2, applies equally to the 1968 table.

Certain observations of some importance, resulting from a comparison of these two tables, are given in the next chapter.

PART 3 (Addendum to Chapter 4)

1969

This chapter concludes with some extracts from the farm report for the operating year of 1969.

The decision to bring the original Haughley Experiment to an end was taken before the 1969 report was available. It is doubtful if its earlier issue would have altered that decision (see Chapter 5) but it would certainly have made untenable some of the arguments on which the changes were pressed in some quarters.

Table 8

CROP	No. of yrs.	ORGANIC Mean cwt./acre	ORGANIC 1968 cwt./acre	MIXED Mean cwt./acre	MIXED 1968 cwt./acre	STOCKLESS Mean cwt./acre	STOCKLESS 1968 cwt./acre	COST per acre FERTILIZER O	COST per acre FERTILIZER M	COST per acre FERTILIZER S	COST per acre SPRAY O	COST per acre SPRAY M	COST per acre SPRAY S
Wheat (Atle)	16	23·7	17·3	29·0	29·6	26·1	32·4	—	158/5	127/7	—	8/–	6/6
Spring Oats (S.147)	7	21·5	28·5	29·2	26·9	—	—	—	108/–	—	—	8/–	—
Barley (Rika)	12	24·5	30·5	32·3	32·6	31·6	36·1	—	109/5	99/10	—	8/–	6/6
Barley (Rika)	12	—	—	—	—	25·8	34·6	—	—	92/8	—	8/–	6/6
Beans	16	13·8	19·3	15·8	24·8	17·7	19·8	—	173/10	115/6	—	75/–	100/–
Peas	16	9·9	6·5	8·6	4·3	—	—	—	135/3	—	—	—	—
Kale	13	312·0	318·0	495·0	766·0	—	—	—	305/3	—	—	—	—
Mangolds	6	313·0	—	506·0	—	—	—	—	—	—	—	—	—
Arable Silage	13	143·1	148·5	216·6	287·5	—	—	—	135/7	—	—	—	—
Lucerne Silage	5	104·7	—	149·1	—	—	—	—	—	—	—	—	—
Lucerne Hay	6	26·3	11·2[1]	32·1	20·0[2]	—	—	—	157/3	—	—	—	—
Silage	5	63·9	—	272·3	145·0	—	—	—	82/–	—	—	—	—
Ley Hay	13	34·8	34·9	45·2	67·3	—	—	—	82/–	—	—	—	—
Sugar Beet	16	—	—	—	—	245·5	—	—	—	228/10	—	—	100/–

[1] Mean of two cuts. [2] Mean of three cuts.

Wheat

'All sections were sown to Atle. This is the twenty-first year of seed obtained from the original stock. Drilling took place on all sections between 15 and 23 October 1968. Germination was good and there was a substantial plant on all sections. On the Organic section a blue tint in the leaf was noticed in March, but this disappeared after a heavy rolling.

Barley

'All crops grew reasonably well, but unfortunately the rains of July lodged the corn on the "M" and "S" sections.

Fertilizer Omission

'In the "M" and "S" sections a strip in each field of wheat was left unfertilized. These were always behind in growth, and poorer in colour, compared with the rest of the field. In the "S" section in particular it was noticeable, despite spraying for weeds, how heavy the weed growth was on the unfertilized section, presumably due to the lack of smothering by a heavy crop. Photographs were taken. The usual mildew attacked all crops.

'In the case of barley, where a strip was left unfertilized on both the "M" and "S" sections (Sawpit-M and Road-S), results were similar to those described for wheat.

'This is the second successive year in which fertilizer has been omitted on the same strip in these fields.

'Loose smut was common in barley on the "M" and "S" sections, despite the use of seed dressing, but was not found on the "O" section where no dressing had been used.'

Beans

'On the "O" and "M" sections Tic beans, derived from the original stock of 1948, were sown. It had been decided last year that owing to the difficulties encountered in harvesting peas satisfactorily, and the resulting low yields, it would be better to grow beans only, instead of having half the field down to peas as in previous years.

'The beans are all winter sown and were drilled in October 1968. There was a good take on all sections. Later on in the season bean aphid attacked both sections heavily, but no sprays were used on either section and the beans grew away from the attack with apparently little subsequent damage. A good harvest was obtained.'

Oats for Silage

'Silage cutting took place during the second week in July on both the "O" and "M" sections.

'It is interesting to note that the small tanks of about 150 gallon capacity which are built to receive effluent from the "O" and "M" silage, were emptied 5½ times on the "O" section against 9 times on the "M". The total amount of silage cut and stacked was 68¼ tons on the "O" and 77 tons on the "M".

Fodder and Root Crops

'The mangolds on both sections were attacked by aphid (greenfly). Spraying with Roger "E" gained control on the "M" section. "Pidero", a mixture of derris and pyrethrum, was resorted to on the "O" section. It was felt that at the time, without the use of a spray, the crop would be lost completely. At a later stage the heavily infested mangolds, some of which had not been sprayed because it was felt to be a waste of time, recovered very effectively and produced a second growth of dark healthy leaves. The total crop yield on the "O" side was, however, poor.

'Again it was noticeable that the heavy woody stalks of the "M" section kale were, to a great extent, left by the cattle, only to cause a problem when it came to ploughing and cultivating during the winter.

'The "O" section kale which is always less robust in appearance (this year particularly poor), but has a more even ratio of stalk to leaf, was entirely consumed by the stock.'

(*It will be noted that this observation confirmed earlier findings. There seems little advantage to be had from forcing an increase in the bulk yield of a fodder crop, if the cattle subsequently refuse to eat it!*)

Grazing

'Full records have been maintained of grazing on each field.

'Again the increase in milk on the "O" section was greater by 15 per cent than the increase on the "M" section when the cattle were put out to grass. To the eye the "M" section was supplying much more grazing.'

(*This also confirms earlier findings (see p.* 233).)

Dairy Herd

'Milk production has improved slightly this year, and although our average is still below the East Anglian average for the breed, the gap has been narrowed. The Organic herd regained its production over the Mixed herd this year. [See Table 6 on p. 246. Also Table 9 on p. 254.]

'There have been a number of cases of infertility this year which caused concern, particularly as the number of cattle returning to the bull did not follow the normal twenty-one-day cycle but was very erratic, with some instances of a return after eight or nine weeks' apparently successful conception. The complete herd was examined by a Senior Veterinary Officer from the Ministry who established that there was no infectious cause for the trouble. Advice was then sought from the School of Veterinary Medicine at Cambridge University and after a very diligent look at the rations by their nutritional expert, it was suggested that the level of cereals on both sections should be increased and that an additional source of phosphorus should be introduced to the Organic ration.

'The cereal ration on the Mixed section was increased by as much as possible but this could not be done on the Organic section without the risk of running out of cereals before the 1970 harvest. However, steamed bonemeal was introduced to the Organic ration at the rate of 28 lb./ton, to raise the phosphate level. The trouble appears to have abated but has resulted in the culling of one cow and one heifer in the Organic herd and four cows and one heifer in the Mixed herd.'

(*The fact that so many more animals had to be culled from the Mixed herd than from the Organic suggests that some other factors were operating besides alleged phosphate deficiency since the 'M' herd had always received bone flour in its rations.*)

Poultry

'Egg production from both flocks followed the usual trend with the "O" flock laying slightly more per bird than the "M" flock.'

(*During the 1970–1 fowl-pest epidemic the Mixed flock caught the*

RELATIONSHIP BETWEEN FOOD INTAKE AND MILK PRODUCTION—1969

TABLE 9

	MEAN 1961–8				PRODUCTION 1969			
	Organic	*Mixed*	*Difference*		*Organic*	*Mixed*	*Difference*	
			Organic higher	*Mixed higher*			*Organic higher*	*Mixed higher*
Total cow-days	4,887·0	5,304·0	—	417·0	5,250·0	5,537·0	—	287·0
Total concentrates (lb.)	13,376·0	16,045·0	—	2,669·0	21,698·0	19,570·0	2,128·0	—
Total concentrates per cow-day (oz.)	43·0	48·0	—	5·0	65·0	56·0	9·0	—
Milking cow-days	3,656·0	4,263·0	—	607·0	5,063·0	4,421·0	642·0	—
Concentrates fed to milkers (lb.)	9,974·0	12,950·0	—	2,976·0	19,769·0	18,081·0	1,688·0	—
Concentrates per milking cow-days (oz.)	43·0	49·0	—	6·0	62·0	65·0	—	3·0
Total milk (lb.)	78,659·0	75,256·0	6,592·0	—	94,550·0	81,285·0	13,265·0	—
Milk per milking cow-day (lb.)	21·4	18·4	3·0	—	18·6	18·3	0·3	—
Milk per acre (lb.)	1,000·5	1,005·5	—	5·0	1,251·7	1,069·5	182·2	—
Milk per lb. concentrates fed to milkers (lb.)	8·4	5·9	2·5	—	4·7	4·5	0·2	—
Milk per lb. total concentrates fed (lb.)	6·1	4·8	1·3	—	4·3	4·1	0·2	—

SUMMARY OF CROP YIELDS—1969

TABLE 10

CROP	No. of years	ORGANIC Mean cwt./acre	ORGANIC 1969 cwt./acre	MIXED Mean cwt./acre	MIXED 1969 cwt./acre	STOCKLESS Mean cwt./acre	STOCKLESS 1969 cwt./acre	COST per acre FERTILIZER O	M	S	COMMENTS
Wheat	17	23·3	21·9	29·0	29·4	26·4	30·0	—	176/–	154/9	
Wheat	—	—	—	—	—	—	26·1	—	—	154/9	2nd succ. year
Spring Oats	8	22·5	22·9	28·9	27·6	—	—	—	106/–	—	
Barley	13	24·9	27·0	32·3	35·5	31·9	28·9	—	106/–	111/8	
Barley			12·2								
Barley	13	—	—	—	—	26·4	26·2	—	106/–	111/8	2nd succ. year
Barley	—	—	—	—	—	—	23·3	—	—	111/8	3rd succ. year
Beans	17	14·1	23·7	16·3	26·6	—	—	—	96/6	—	
Kale	14	312·0	216·0	514·0	778·0	—	—	—	291/8	—	
Mangolds	6	313·0	352·0	506·0	800·0	—	—	—	291/8	—	
Arable Silage	14	143·5	164·4	221·6	253·3	—	—	—	159/–	—	
Lucerne Ley	7	24·1	12·5[1]	27·8	30·2[1]	—	—	—	—	—	
Ley Silage	6	—	—	251·0	200·0	—	—	—	—	—	
P.P. Hay	—	—	—	—	10·0	—	—	—	—	—	
Ley Hay	14	34·8	35·4	46·7	36·8[2]	—	—	—	—	—	

[1] Mean of two cuts.

[2] Mean of three cuts.

disease, and were all lost. For many weeks it looked as though the Organic flock would escape, but in the end it too succumbed and was lost.)

Sheep

'A fairly good year for lambing once again with an average of 1·6 lambs per ewe on the "O" section and 1·3 lambs per ewe on the "M" section. The young lambs sold well in June at the local lamb auction.'

(*In 1970 both flocks were regrettably sold.*)

5

Ending the Experiment

From 1947 to 1967 the Haughley Experiment was sponsored by the Soil Association through its world membership; final authority for policy and finance was vested in the Council, a body elected democratically by the whole membership.

The Council appointed the Research Director, and laid down farm policy in consultation with various consultants, including the Research Director, who visited the farm weekly.

Responsibility for carrying out the agreed policy was delegated by the Council to the directors of Haughley Research Farms Ltd., a subsidiary trading company with a share capital wholly owned by the Soil Association.

Day-to-day operations were in the charge of a farm manager who worked in close liaison with the Research Director and who was responsible to the company board of directors for detailed reports and records.

Fundamental Research is notoriously costly and a project such as the Haughley Experiment, to fulfil its potential, needed facilities and resources on the scale of a university rather than that of a small voluntary society without grant aid. It is not surprising, therefore, that the progress of the experiment was punctuated by financial crises. The surprising thing is that, in spite of this, so much of lasting value was achieved (see Chapter 6). Throughout the twenty years of Soil Association sponsorship only the enthusiasm and generosity of members enabled the experiment to survive recurring threats of bankruptcy.

One of the worst of these crises occurred in 1964–5 when, not only did the future of the experiment seem doomed beyond reprieve, but the very existence of the Soil Association itself was gravely threatened.

This time the majority of the Council decided that the crisis called for crisis action. This resulted, among other things, in a break in continuity of management policy and of the research programme, and led to the appointment of Mr. Douglas Campbell[1] to the post of what amounted to generalissimo. He was first asked to take charge of the farming operations as combined Farm Manager and Research Director. Shortly afterwards he was also made administrator of the Soil Association, as well as chief (virtually sole) negotiator in all fund-raising efforts. In retrospect this was clearly more than any one man should have been asked to do. Mr. Campbell shouldered the burden manfully, but the unquestioning resolution with which he did so also made it only too easy for the Council to shelve certain responsibilities which—or so it seems to the writer—ought never to have been delegated.

Not long after this appointment a miracle occurred. In 1967 Mr. and Mrs. Jack Pye, founders of the Jack and Mary Pye Charitable Trust, decided, both as individuals and through their Trust, to save the research farms and to back both the experiment and the Soil Association with generous grants for a minimum period of three years. They acquired the freehold of the farms from the then owners, who were planning to evict, and gave Haughley Research Farms Ltd. security of tenure in a proper lease. They next undertook an extensive programme of capital improvements in buildings and roads, and built and equipped a highly sophisticated small-animal research unit.

Development of the Haughley Experiment seemed set fair. The very fact, however, that for the first time there was a prospect of being able to undertake long-term planning, with an assured income, disclosed a deep cleavage of opinion and objective within the Soil Association Council.

Many of the original members of Council had retired or died and been replaced by a younger generation. Many of these, and a few older members as well, had come to the conclusion that world changes since 1947 (including the new surge in conservation movements; the wide-scale awakening to the environmental crisis; the changing attitudes among agricultural scientists to the organic approach, and, not least, the increasing public demand for 'whole food') had so altered the situation that it was now far more important for the Soil Association to use the farms to demonstrate productive Organic farming, and to undertaking *applied* research (for the benefit of farmer members) than to continue to engage in increasingly costly basic research.

Those holding these views, both inside and outside the Council, naturally wanted to see production stepped up on the Organic section, and they urged a change to modern varieties of cereals and the

[1] See Appendix 1.

abandonment of the closed-cycle concept, on the grounds of 'evidence' they believed recent farm results to provide (notably as shown in the 1968 report) that the Organic section was 'running down'.

Those holding the opposite view argued that other institutions, and existing successful commercial Organic farms, were better equipped than Haughley, for applied research and demonstration. More important still, they maintained that the basic research material which the Haughley Experiment had built up over twenty years was unique in the world, and was only now ripe for scientific investigation. To change the whole basis of the research at this stage would be to throw away a priceless asset and, with it, the growing interest among a wide variety of scientists eager to make use of the material in their own specialisms.

To the alarm concerning the alleged 'run-down' their attitude was: 'What of it? It is part of the experiment to find out at what point the Organic section does start to run down, if it does, and to investigate and interpret the factors responsible.'

Dr. K. Mellanby, Chairman of the Haughley Research Advisory Committee,[1] wrote, in the April 1969 issue of the Journal of the Soil Association (p. 336):

'We are all agreed that no fundamental changes should be made in the next few years and only then after a very thorough investigation. . . . I think that all organic farmers would agree that with the knowledge they have today, they could suggest ways in which to improve the Organic section so as to get larger yields and to make the whole output more commercial. This, however, must not persuade us to make any alterations until we've abstracted all the scientific information possible from the present set-up.'

While the argument continued, both in the Council and in the Research Advisory Committee, Mr. J. F. Ward[1] was asked to study all the farm and analytical reports for the period 1952–65 and to prepare a summary, adding his own evaluation of the result in terms of evidence enabling a deduction on trends to be made. It was presumably hoped that this would settle the question of run-down one way or the other.

A long document resulted. It makes interesting reading and, like all other reports referred to in this account, is available for study.

On the question of a decline on the Organic section no very clear picture emerged. A section entitled 'Some General Aspects Accruing from the Three Sectional Treatments' starts off with the following:

[1] See Appendix 1.

'Considerable evidence, mostly visual, has been recorded that the soils and crops of the three sections are behaving in different ways.

'The workability of the soil can be modified considerably by tillage and manurial operations. During soil sampling the easy working of certain fields is reflected in the ease or difficulty of taking the samples. Tillage operations, such as ploughing, seeding, etc., and tractor consumption of fuel are all of use in demonstrating the variations in the workability of the soil. Observations, such as these, point to the better texture and hence the greater ease of working the fields with high organic content and in this respect to the superiority of the fields in the Organic section over the other two sections. There is mounting evidence that the fields on the Stockless section are in general becoming more intractable.

'The palatability of the herbage may be a better guide to its nutritive value [quality] than present analytical methods. For instance, it has frequently been observed that when the cattle are first turned out in the spring on to a Mixed ley they first graze the headlands and ditch banks, where the plough does not pass, and, if a strip has been left unfertilized, they will eat that off before consuming the fertilized area.'

Later on in his report he presented the pros and cons for a decline, thus:

Factors Pointing to a Decline (on the Organic section)

'(a) SOIL NUTRIENTS: Potash low and possibly falling.

'(b) CROP YIELDS: No increase in yield of wheat, barley or beans, a drop in the yield of silage and, in some years, a failure of the leys to produce a hay crop is descriptive of the Organic section. On the other two sections, Mixed and Stockless, there is a general increase in crop yields.

'(c) CROP ANALYSIS: Content of potash in the wheat grain is falling more rapidly than on the Mixed section.

'(d) MILK YIELD: That of the Mixed section is catching up the yield of the Organic section calculated on an acreage basis.

'(e) MILK ANALYSIS: Increase in total solids more marked in the Mixed milk. Fall in content of nitrogen and phosphates more marked in the Organic milk. Butterfat lower on the Organic and falling more rapidly.'

Factors Against a Decline in Fertility

'(a) The Organic herd produces more milk from less food.

'(b) Evidence that withholding fertilizers from the Mixed section causes a reduction in yield to below that of the corresponding Organic crop, leading to the corollary that maintenance of the yield on the Mixed section is dependent upon the application of fertilizers and not upon the inherent fertility of the soil. Further investigations are required before this can be accepted as a factor.'

Mr. Ward could have added more 'factors' drawn from his own report, for, even without the 1969 figures and observations (given in the last chapter), it is clear that the Organic section production remained quite remarkably steady throughout the period under review. Again and again he uses the phrase, 'No change on the Organic'. Even where a fall in one or other nutrient element is recorded, the opposite seems to have occurred with some other element. Study of the 1968 crop-yield record (Table 8, p. 250), which was actually used by some as evidence of Organic section run-down, shows no less than six Organic crops that gave a *higher* yield that year than the mean for the previous years. This was true even of kale, the poor yield of which was thought by some to 'prove' a decline.

It seems to the writer that what the figures and comments in Mr. Ward's summary provide, is evidence that the Mixed and Stockless sections had been stepped up (by increased fertilizer application?), but that the Organic section had become, or was becoming, *stabilized*; *a very important finding* if true, not sufficiently brought out by Mr. Ward in the final paragraph of his list of factors in the reference to 'inherent fertility'.

Only the statements on milk (yields, analysis, and butterfat) give grounds to suggest the possibility of a decline, but the 1969 figures (see p. 246) show a comeback by the Organic herd. The 1969 report confirms this, and in other respects fails to support the run-down theory. For example, the livestock figures show that at the end of 1969 the Organic section was carrying eight more head of cattle and twelve more sheep than at the end of 1961, and one more cow and five more sheep than in 1964.[1]

The foregoing has been included merely for the purpose of putting the record straight, for it was not, in the end, the presence or absence of such evidence that proved the determining factor in bringing about

[1] See Appendix 2, p. 302.

the changes and, with them, the end of *the* Haughley Experiment. It was a combination of the majority wish that the Organic section should become a convincing example of 'good' (meaning productive) Organic farming and the assumption (not altogether unjustified in view of past history) that no source of large-scale backing for basic research would be forthcoming beyond the initial period promised by the Pye Trust.

The issue aroused strong feelings on both sides. At the height of the controversy Mr. Campbell invited selected members of the Council, and certain others, to form a working party at a weekend conference to be held at the farm, with the object of producing a considered recommendation for the Council. Only two of those who understood the basis of the Experiment, and the significance of the closed cycle, received invitations (the director of research and the founder were deliberately excluded).

As was to be expected under these circumstances, the conference produced a majority report recommending basic changes in the research design and objective.

There is little doubt that this weekend was the main single event that tipped the scale. At the subsequent Council meeting the Working Party recommendation was accepted by a majority vote. (A proposal to submit the issue to the whole membership by referendum, as had been done in previous crises, was rejected as being unconstitutional.)

Once the die was cast, the minority conceded defeat and looked for ways in which it might be possible to co-operate with the majority in making a success of the new approach, because all concerned, whatever their opinion on the controversial main issue, were equally keen that, in replacing the Haughley Experiment by the new objective, the unique research material that had accumulated (see next chapter) should not be lost.

Then, in 1970, strange things began to happen. It was as though with a disintegration of a common will and aim in the Council a *general* disintegration set in, accompanied by the worst financial crisis in the organization's whole chequered career.

This revealed, with painful clarity, that under prevailing conditions the Soil Association was quite unable any longer to shoulder the responsibility for financing the farms. Their assets would have to be sold up, or the Soil Association itself become bankrupt.

The Pye Charitable Trust once more stepped in, offering to take over the farms, lock, stock and barrel; to keep the three sections as separate Organic, Mixed and Stockless units; to retain the two dairy herds, and to carry out research work in line with the Soil Association's new objectives.

Thus a new chapter has opened, both for the farms—now called the Pye Research Centre—and for the Soil Association, which, after a year of marking time, devoted to recovery and re-organization, is now (1973) launching out under new administration.

As these new chapters unfold, it is hoped that the future may bring to the fore many things of common interest and enlightenment to both the Pye Research Centre and the Soil Association, which are now separate undertakings under separate management.

6

The Achievements and Some Unanswered Questions

I assembled the first five chapters of Part II in an editorial capacity; they were compiled mainly for Soil Association members, but also for any other readers interested in the subject.

I write this chapter in my capacity as co-founder of the Haughley Experiment, and therefore do so in the first person.

It is addressed primarily to some unknown future investigator who, so is my hope, will one day study the history of this pioneer experiment in full awareness that what the science of ecology is faced with is 'functional relationships'; these are to be seen in the patterns of action of living entities rather than in mere material quantitation.

When the day comes, I hope further, that my unknown ecologist will be able to arrange for the threads of the experiment (with their many loose ends) to be picked up again, repeated and developed, and this time carried on to their logical conclusion.

With such a person in mind it is my purpose, in this chapter, to indicate what the experiment actually achieved in its short life, and to draw attention to the number of questions, many of them unexpected, which the experiment threw up but which were never answered.

First, however, I have one remaining editorial duty to perform; namely to anticipate and answer, a possible criticism.

It will be realized that in order to reduce the very large quantity of written matter and figures (that twenty years of research gave rise to) not only into a document of this size, but also (I hope) in readable form, only a very small selection from the wide choice of available material could be made.

It may be said, by those who are familiar with the originals, or become so in the future, that I have chosen to select more findings that favour the Organic section than the reverse. To the degree that this criticism is valid, the reason for it is that, from the orthodox point of view, after twenty-five years of operating the three sections, *all* of which exported produce but only *one* of which (the organic) *imported nothing, there ought not to have been any findings at all that favoured that section.*

The fact that there were so large a number, quantitatively as well as qualitatively, completely justifies the view that the original design of the experiment was capable of producing information which other methods were not.

The unanswered questions, as will be seen, arose *because of the closed cycle*, not in spite of it.

Many people made real sacrifices in the hope of keeping the experiment going long enough for something of scientific value to emerge. They, as much as any future investigator, are entitled to know how far they succeeded. So what *was* achieved?

I

First, I think, must be put the accumulation of the unique research material that has resulted from twenty-five years of careful records covering the three different systems of land use.

This still has value not only for the original aims of the Haughley Experiment, but for various world-wide research projects for which no such material is available anywhere else in the world.

There are at least four categories of such material.

First is the soil itself. Originally uniform in character over the whole area[1] it has been possible to watch and record, both from field operations and laboratory analyses, the progressive changes which the three systems brought about—in humus content, soil aggregates; water-holding capacity; porosity; ease of working, etc.—in fact—tilth.

After twenty years these changes were such that, when they were inspected by Ministry of Agriculture representatives, what they found influenced the government report on soil structure issued in 1971.

Second is the existence of an area of land (the Organic section) large enough to be a rotational working farm-unit, on which there are records that can guarantee that no insecticides, or herbicides, or other agro-chemicals have been used for at least two decades.

When Dr. K. Mellanby, as Director of Monkswood Experimental Station, was commissioned by the Ministry to make a report on the

[1] Except for a few minor differences in a few fields, and these were represented on all three sections.

effect of agro-chemicals on wild life, including insect populations, he rightly pointed out that before this could be done it was necessary to study the balance of such populations where no such chemicals were used. He had to ask 'Haughley's' permission to make this study on the Organic section since *no other area of working farmland now existed*[1] that fulfilled the requirements.

A *third* category of unique research material is the cereal and pulse grain. This was originally undifferentiated, bought seed when the initial sowing on all three sections was made, but was, thereafter, grown, and sown, year after year only on its own section.

By 1970 this system had produced: nineteenth-generation wheat and beans on each section, thirteenth-generation barley on each section, and tenth-generation oats on each of the Organic and Mixed sections. When the experiment came to an end several cwt. of each of the above were put in store and, at the time of writing, are still available for research projects. This material, too, is obtainable nowhere else.

A *fourth* example is the third- and fourth-generation cows of the two closely related Guernsey herds, nurtured from the time of their foundation-cow ancestors, only on food from their own section.

In listing the above items of 'unique research material' which the Experiment produced, it must be pointed out that, although the present tense has been used, not all this material is still available. What remains (though somewhat blurred by post-1965 changes), at the time of writing, still has unique features, namely the soil of the Organic section, and the home-bred, home-fed herd of cattle, with the recorded data of both—but these are no longer part of a diversified and integrated ecological food-chain, functioning *as a whole*. Many of the original components are missing altogether (such as sheep and poultry), and the importation of new seed and organic soil adjuvants means that the effect of the nutrition cycle through successive generations, of plants as well as animals, can no longer be observed.

II

I turn now to the positive results that contributed to scientific knowledge or to a better understanding of the ecology of soil, plant and animal.

(a) Of these, perhaps *the* most important thing to come out of the Haughley Experiment, because so conclusive and, surprisingly, hitherto unsuspected, were the findings that *the levels of available minerals*

[1] Author's italics.

fluctuate according to the seasons[1] (see Graph 1, p. 217), and that in an Organic soil medium there is a symbiotic relationship between the plant and the flora and fauna of the soil whereby the latter release minerals at the root tip for uptake into the plant at the appropriate times, and in minute continuous quantities. This fluctuation was far more marked on the Organic sections than on the others.

This finding would have been obscured if imported material had been applied to the Organic section. The fact that it was so clear cut was an *outcome of the closed cycle.*

(b) Another important finding for which we have the closed cycle to thank, is the dependence of the Mixed farm on the artificial fertilizers applied over the years, and the dramatic effect which followed the strip experiments where no fertilizer was applied.

I would here draw the reader's attention, once more, to Table 5 on p. 240 and to the corroboratory comment by J. Ward on p. 261.

The lack of a similar dependence (on outside aids) on the Organic section indicates the self-supporting nature of the biological fertility which was becoming established on that section. This may well be connected with:

(c) The somewhat puzzling finding that the average humus level, over the experimental period, was consistently higher on the Organic section than the Mixed, despite the fact that the latter had regularly returned to it as much Organic matter as the Organic section, and indeed if we accept that the Mixed leys will produce more crop, then the Mixed section has received greater quantities.

(d) The fact that humus brings about better soil aggregates which produces better drainage, better aeration and better nitrogen fixation (see p. 276) may help to explain the better utilization which the Organic-section crops appeared to make of their soil environment.

A consistent finding, particularly with autumn-sown cereals, was a visual observation of an apparently much delayed growth in the early stages on the Organic section. Further examination, however, showed that in this initial period the plant in an Organic environment is 'concentrating' (if I may so put it) in establishing a vigorous root system. Having done so, but not before, it is ready to make top growth (i.e. the *behaviour pattern* of growth is quite different to that of plants growing in a chemical or 'mixed' environment).

This interpretation is supported by the fact that before the end of the growing season the Organic crops caught up the others, and,

[1] Following Dr. Milton's publication of this finding, his work was repeated and confirmed at university level.

as we have seen, remained able to look after themselves. It would seem then, that in an Organic environment, the initial period when root development is occurring can give the appearance of delayed crop growth but because of this early natural attention to the build-up of an effective root system, better plants can subsequently be obtained.

It is important to remember that Organic farming is not merely a matter of the type and quantity of materials supplied to the soil. It is also a matter of management, which includes a lively awareness of the natural changing conditions in the environment.

Among these are, of course, temperature and moisture. An Organic system is much more dependent on these than other systems. For example, with a late spring there will be a late awakening of the living elements in the soil and therefore a delay in the growth of the plant.

(e) This was very clearly demonstrated on the leys of the two sections. In every early spring, during the experimental period, the Organic leys gave the earliest bite, in spite of no fertilizer treatment. In years with late, cold springs (and being East Anglia these were in the majority) an earlier bite could be induced in the Mixed leys by quick-acting nitrogen top dressings. But that this does not necessarily produce a better quality in the herbage is shown not only by the response of the two herds in milk output from their respective pasturage (a difference over twenty years of around 15 per cent, see also Graph 5, p. 233) but also by the obviously greater palatability of the Organic leys, and the greater contentment of the Organic cows grazing thereon.

In making these various comparisons between the Organic and Mixed leys, other factors were, of course, involved besides the presence or absence of fertilizers.

Methods of mobilizing the natural minerals including trace minerals, in soil, constitute some of the most important parts of Organic farming management. In pastures these methods include not only techniques to encourage proliferation among the soil populations (such as surface-soil aeration, and provision of Organic food) but also the choice of ley mixtures containing deep-rooting grasses, legumes and herbs. Such deep-rooting plants, from trees to plantains, draw up minerals and trace minerals from the lower levels of the soil, and redistribute them in the top layer for use by the more shallow rooting crop-plants.

The mixtures used at Haughley are given in Appendix 2. Their efficacy in this respect was demonstrated in one of the very early findings when dried seaweed, on an *ad-lib* basis, was first offered to Organic grazing animals in summer, as a trace-mineral supplement.

It is a characteristic (in fact the purpose) of such, not particularly palatable, supplements that (when fed *ad-lib*) they are consumed whenever the animal's own instinct leads it to correct some deficiency in its diet, and not otherwise.

(f) Throughout the summer that gave rise to the following finding, the whole of the Organic herd, of all age groups, had two adjoining leys for their summer rotational grazing. One contained the range of deep-rooting herbs afterwards adopted as standard. The other was without them, containing only a mixture of different grasses and clovers.

Both leys were, visually, equally productive. Access to dried seaweed was made available on both.

The herd was moved regularly back and forth from one to the other to permit of the necessary rest-periods which all good pasture-management requires.

During each period that the cattle spent on the ley lacking the herbs, the seaweed was eagerly consumed, and had to be regularly replenished. Whenever they were feeding on the ley which included the herbs (or weeds as they are often considered) the seaweed was scarcely touched.

This episode illustrates clearly (i) the value of a herbal ley far outweighs any disadvantage due to its lack of bulk; and (ii) the importance of studying animal choice. The cow often knows better than the chemist what is good for her. Scientists in animal husbandry might achieve more if they consulted the cow more often.

(g) Another finding worthy to be included in this list, because its significance was, once more, revealed through the existence of the Organic section closed cycle, is that the chemical composition of the Organic crops, which received nothing, never averaged lower levels than those of the other two sections receiving fertilizers.

The reverse side of this finding is equally interesting. It reveals a great wastefulness in normal fertilizer applications, for the analyses of the Mixed and Stockless crops showed that only minute quantities of the amounts of soluble minerals applied as fertilizers were recovered in the crops harvested. This underlines the fact that random plot experiments of the orthodox type have been concerned only in measuring apparent increase in yield without taking into consideration the true cost, the wastefulness and the support given by the subsidy system. (True biological economy is not the concern of such experiments.)

(h) The finding that variations in soil nutrient levels are affected by different categories of crops in a rotation, is a discovery worthy of further research. The contrast, it will be recalled, was particularly marked between leys and annual arable crops (see again graphs on pp. 227 and 228).

III

It must not be supposed that the observations and findings listed above all occurred at once, or that their significance was apparent in a short period of time. Nevertheless, it must have become obvious to the reader of the foregoing, that a number of unanswered questions of considerable potential importance have been thrown up.

This in itself was a remarkable achievement in so short a time, and with such slender resources.

It must be emphasized that the work had only just begun before it ended. Only by the time the experiment was abruptly changed had a sufficient number of pointers emerged to show the way in which a future investigation of this kind could most fruitfully be conducted.

In the next (and final) chapter, contributed by Dr. Milton, he has commented on, and where necessary amplified, these pointers, and has set out some guide-lines to indicate the ways in which a future, far-seeing ecologist might proceed with a fundamental biological experiment to arrive at correct interpretations.

By way of preparation for this, I end this chapter by running through the findings (in summary) described above in such a way as to bring out the unsolved questions they pose. (Page reference is given to assist back checking for greater detail.)

Page ref. of Finding No.

(a) p. 266 The fluctuation of levels of *available* minerals in the soil according to the time of year. Maximum release coinciding with time of maximum plant demand.

QUESTION:

What are the biological sequences which, when undisturbed, hold plant nutrients locked up when not required (so preventing leaching) and release them when needed?

Recent microbiological investigation of the rhizosphere has thrown some light on the symbiosis between the plant and the flora and fauna of the soil, which is almost certainly involved, but so far *none at all* on why this relationship works less efficiently when soluble man-made nutrients are added to the soil. (See again Graph 1, p. 217).

This needs to be discovered.

(b) p. 267 The dependence of the Mixed section on the fertilizers applied over the years, against a self-supporting biological fertility on the Organic.

QUESTIONS:

(i) *What are the biological processes which are depressed by fertilizers even in the presence of as much, or more, organic residues?*

(ii) *Does the addition of the additives, i.e. 'foreign' material, cause the plant to alter its mode or pattern of metabolism for utilization of the material presented to it?*

(iii) *Is this change operative only in the immediate presence of the available food; or, does any such change, in food available, confer a permanent or semi-permanent change in its metabolic pattern of action?* In humans there is a tendency in the presence of insult or injury to revert to basic patterns of metabolism learnt in infancy. This may be a general biological principle for all living organisms.

(c) p. 267 Average humus level higher on the Organic section than the Mixed, despite the level of organic treatment being *no* higher.

QUESTION:

Is the build-up of soil microflora and fauna greater where (i) the organic return to the soil is not accompanied by chemical additives and/or *(ii) where the organic residues are composted before return to the soil? As was the rule in the case of the Organic section.* (On the Mixed section they were applied as partly rolled farmyard manure.)

(d) p. 267 The Organic plant's hold-back in top growth pending the development of a self-supporting root system.

QUESTION:

Is this evidence of a different action-pattern in the organism, arising from a qualitative difference in the nurture supplied by its environment?

(e) p. 268 More milk from less food from the Organic herd than the Mixed. Particularly the markedly higher production of the Organic herd when out at grass, on comparatively sparse leys, than that of the Mixed herd on apparently much lusher and more productive ones. Also the greater contentment of the Organic cows.

QUESTIONS:

This finding poses several questions:

(i) *How much of the greater bulk of the Mixed pasture is not food at all but water?* (See figures on p. 340.)

(ii) *Is pasture, grown organically (a positive approach) or one grown without NPK fertilizers (a negative approach) not only more palatable (which it demonstrably is) but more nutritious?* or

(iii) *Is its utilization by the cow, in the metabolic processes in the rumen different; and more efficient?*

(iv) *Is the change in the metabolism effected, one which, for example, changes the nature of the flora of the gut; which in turn modifies the metabolism for digestion by the hosts?*

Investigation along these lines evokes an exploration which is essentially an ecological one.[1]

(f) p. 269 The connection, imparted by the animals themselves, between the function of deep-rooting land herbs, and sea plants.

QUESTION:

Is this an indication that much more ecological study should be made of the animal instinct that governs its choice of food, when such choice is available? (See work done by Dr. Albrecht, Appendix 11.)

(g) p. 269 Concentration of minerals, as shown by analysis, is not less in the Organic crops than in the others where fertilizers have been applied.

QUESTION:

What, if any, is the correlation between application of fertilizers and their recovery in crops?

(h) p. 269 Levels of soil nutrient availability are affected by the *type* of crop growing in that soil.

QUESTION:

The question posed by this finding brings us back full circle to the first finding, (a) p. 266 (see p. 270). *Together they*

[1] 'In a study of quality and health many of the recognized manifestations of health defy measurement. We must not forget that "the biological proof of the pudding is not in its constituents—but in the eating". Hence the critical factor in the study of "living" (i.e. health) is that of utilization. It lies in *how* the particular organism uses the material means available.' (Dr. G. Scott-Williamson.)

> *indicate that a complex, and little understood, functional interrelationship exists between soil minerals, soil micro-flora and fauna; the plants themselves; the influence of animals on the plants, and seasonal weather conditions.*

A true understanding of plant nutrition can never be arrived at if any of these factors is overlooked.

Perhaps the achievements of the Haughley Experiment can best be summed up as having demonstrated quite clearly the need for a fully ecological investigation of the 'wholeness' of the soil–plant–animal–man–soil food-chain.

7

The Haughley Experiment—Some Unsolved Problems, Their Significance and Suggested Methods for Investigation

by Dr. R. F. MILTON

The foregoing description of the Haughley Experiment clearly indicates that at the time it was abandoned it had proceeded to a stage full of potential for the establishment of considerable understanding of the life-cycle—soil–plant–animal–man.

The decision to change the layout of the experiment was made as a result of pressure from a majority of Council members who at the time appeared to be convinced that the farms were 'running down' and as such did not represent a good public image to the Organic movement. There was no scientific evidence for the assertion. In fact it was rather to the contrary. It is true that the Haughley Farms as operated within the limitations of the Closed Cycle Experiment could not compare to the best-run organic farms with regard to productivity, but the latter were not restricted in the importation of organic matter and were able to use newer varieties of higher yielding seed. (In many instances the seeds used at Haughley had been deliberately resown continuously since 1946.)

It was precisely to investigate the feasibility and the effect on soil fertility and plant and animal health when operating without brought-in adjuvants and continuing the seed strains that was the basis of the experiment.

Clearly the majority of Council Members were out of sympathy with or did not understand the underlying implications of the above conception.

Of course, as I have frequently stated, there are many ways in which agricultural experiments can be devised, and the modifications made at Haughley after the decision to break the closed cycle undoubtedly would still have produced much information useful to the Organic movement, had this work been allowed to continue.

As we have very little concept of how the farms are now being managed, comment on the future would be mere conjecture, but it is anticipated that under the Pye Trust the experiment will remain concerned with investigation into organic methods of cultivation.

The cessation of the original Haughley Experiment must be considered as unfortunate in view of the fact that the stage had been reached, as a result of work carried out, for considerable expansion to take place of the investigational programme into realms such as the effect of different cultivations upon the true nutritional value of food.

Furthermore the 'climate' of opinion is now at the point where increasingly it is realized that with expanding world population we cannot much longer tolerate robbing the reserves from one region in order to augment or bolster up the agricultural economy of another more affluently placed.

Thus it has recently been pointed out that in the UK we import annually 700,000 tons of costly potash and phosphate raw materials for fertilizer manufacture and at the same time a greater potential tonnage of fertilizers excreted from the human body are washed away to pollute the rivers and sea. More people are becoming conscious of the fact that the Consumer–Capitalist ethic encourages society's intemperate and non-sustainable depletion of natural resources and many are now making attempts to collaborate with nature and so reverse the process of continuous and inefficient plundering of non-renewable materials.

The analytical work carried out in connection with the Haughley Experiment has shown how wasteful of natural resources is modern commercial farming and how with a closed-cycle technique nutrients are recycled and moreover become available *in situ* provided that an ecological approach is made to the methods of cultivation and farm management.

Thus we have shown that with intensive fertilizer application on a mixed farm only about one-fifth of the chemical applied is recovered in the crop. The rest either finds itself in bound form unavailable to the plant or is leached into the deeper layers to pollute brooks, streams, lakes and rivers.

Furthermore, we have shown how by encouraging the proliferation of the appropriate soil bacteria latent nutrients in the soil are made available at the root tip in concentrations naturally acceptable to the plant constantly throughout the growing period. This is a field for considerable further work in elucidating the mechanisms of symbiosis between the soil organisms and the plant system.

On the Organic side of the Haughley Farms reliance was made on the nitrogen-fixing organisms of the soil to provide the necessary nitrogen for plant use. The pointers from the original experimental work indicate that much valuable information could be obtained by studying more intently the mechanisms involved and so further reduce reliance on chemical nitrogen.

Our Organic balance-sheet indicated that with removal from the closed system of only milk, eggs, wool and culled animals, there is available in the average top soil enough potash and phosphate and trace minerals for about 1,000 years, even without the possibility of replenishment by way of surfacing from the deeper layers of the soil through the agency of deep-rooting plants, shrubs and trees. This is a field for further research over a wider range of crops with emphasis on availability control.

We saw how dependent the success of the Organic method is upon cultivations made appropriately at the optimum moment and more research into this factor is called for.

Likewise, we observed how the awakening of soil life in spring and with it concomitant plant growth was so dependent on soil temperature, and although we assume that this is connected with rate of microbiological activity in the soil, the subject undoubtedly calls for deep investigation.

There was the visual observation that crops grown on land laced with chemical fertilizer appear to become dependent on its continued application, rather like a drug addict. To what extent this observation can be confirmed and the underlying biology thereof requires further study.

Organic-section cows seemed consistently to produce more milk on less fodder than their chemical-section counterparts. This is an important finding which should be followed up and then investigated on other herds operating under other circumstances. The finding calls for an intensive study to discover whether this is linked to nutritional differences in the crops grown on the two sections.

The importance of the humus factor in the soil was brought out in the early work. Its relationship to the formation of soil aggregates which are essential for drainage, moisture control and aeration calls for further work. Soil conditioning and the part played in it by humic substances was brought to light and needs elucidation.

Perhaps we must emphasize that the past work showed the futility of judging the nutrient status of the soil from spot chemical analyses carried out on samples taken at any time of the year. Our pointers would seem to indicate that a much more ecological approach to the assessment of soil fertility is obtained by attempting to measure growth capacity of the organic life in the soil particularly at the root tip. In this connection the work and methods of Dr. Rusch becomes relevant, which amounts to attempts to study the biological changes which are produced in the soil associated with plant growth.

The manner in which to tackle these problems calls for recourse to all available methods of scientific investigation. As newer techniques develop so the field of exploration will become widened. Some proposed methods for investigation were outlined in the memorandum which I prepared for the Soil Association Council before the closing down of the experiment. In this it was pointed out that the work carried out previously was in the nature of probing to find some salient features which would profitably justify wider investigation. These features have largely been high-lighted throughout this report.

It is suggested that the finding of 'more milk from less food' if substantiated must have far-reaching implications. There are numerous ways in which this may be investigated, but the finding does stimulate the question 'Are there nutritional differences in the fodder which relates to the soil on which it is grown?' This might be investigated along the following lines:

Small-animal Feeding Tests Using the Crops from the Various Sections

Some such tests have already been in progress at Haughley for some years now, sponsored by the Pye Trust. We have had no indication to date as to whether any conclusions of any kind have been reached from the work so far carried out.

There are some differences of opinion amongst Soil Association members as to the significance of any findings which might come from the approach now followed in view of the extremely artificial conditions under which the animals are housed. It has been voiced that the diet as fed is too restrictive and that this in itself must invalidate extrapolation of the findings to the other animals.

Furthermore in view of the changes which have taken place in the management of the farms since the demise of the Haughley Experiment, we have no indication that the produce as now fed can be related to that which was available when the observation on milk differences was established. Nevertheless, in my opinion direct animal feeding experiments suitably devised still remains the most likely short-term method to show any changes in well-being, etc., resultant from true nutritional differences in the food grown on differently managed soils, if indeed such exists, and it should be pursued when opportunity allows, possibly in a more realistic manner.

Chemical Composition

During the period that we analysed crops grown on the various sections, we were unable to demonstrate consistent compositional differences with regard to total proteins, oil, carbohydrates, mineral substances and vitamins. It is considered, however, that differences in amino acid pattern might emerge from a quantitative investigation into these substances. With modern amino acid mass analysers this would not be a formidable task.

Variations in Rumen Constituents

The utilization of ingested food by ruminants is dependent upon the composition of the rumen fluid. This in turn is influenced by the composition of the fodder ingested. With cows fistulated it is possible to investigate to a considerable degree the biochemical processes which go on in the rumen and to predict the potential energy absorption therefrom into the animal system. To what extent this is bound up with subtle or even large changes in rumen flora can be concomitantly checked. The accepted view of rumen digestion is that it results in the production of short-chained fatty acids to sustain animal energy requirements. It is not difficult technically to investigate this and changes brought about as a result of diet modification.

Saponin Content of Fodder and its Relationship to Rumen Efficiency

It is well known that bacterial efficiency in the rumen is enhanced by the presence of small amounts of saponins which act as surface-tension reducing agents and thus allow more ready contact between micro-organisms and ingested fodder.

Preliminary work at Haughley indicated that fodder from organic leys containing a diversified range of plants had a definite saponin content which could not be identified in the orthodox rye-grass mixtures which formed the basis of the fodder on the mixed farm. The possibility that this is a factor in the increased utilization of fodder in the organic herd *vis-à-vis* the mixed is a subject worthy of closer investigation.

Physical and Chemical Basis of Palatability and its Relationship to Nutritional Quality of Food

It has been frequently observed that animals search round for certain plants for preferential feeding. This could be a starting point in the design of work to assess palatability factors. The Haughley cattle appeared to be avid feeders on the deep-rooting plants on the leys. If a statistical case was proven, analysis to seek the palatability factor could be rewarding (see p. 269).

Differences in Composition of the Fatty Acids in Plants in Relation to Soil Management

Preliminary work in this field had already been undertaken prior to the end of the Experiment. This work indicated that larger proportions of the saturated types of long-chained fatty acids occurred in crops grown using chemical fertilizers.

Composition of Produce (Milk and Eggs) in Relation to Soil Management

Much analytical work was carried out in this field on Haughley produce and the results obtained are given in the various annual reports. Certain abstracts are reproduced in Appendix 6. Future work could concern attempts to ascertain whether differences exist in (a) amino acids, (b) protein fractions, (c) nucleic acids levels and composition of foods from the two sections.

What is the Difference Between Compost and FYM?

(See p. 271 (c))

The use of compost on the organic section and of farmyard manure on the mixed section has always raised questions as to significance. Little is understood about the effect of composting other than that the result is richer in humic matter and poorer in cellulose content.

Sophisticated analytical technique must be applied here. This is a field for major investigation.

The Depression of Certain Soil Micro-organism Activity on Addition of Chemical Fertilizers

To what extent this occurs and under what conditions needs full investigation (see p. 271 (b)). There is some evidence from past work that this phenomenon is influenced by the laws of mass action. It can be further investigated.

The Mechanics of Bacteriological Release of Plant Nutrients from the Soil Symbiotically with Root Environmental Changes

The observed finding of increased nutrient levels in the soil relative to plant growth allows of considerable scope for investigation, using both chemical and microbiological technique.

In an Organic Soil Top-growth of the Plant tends to be Depressed Pending Increased Development of the Root System

Clearly this is a difficult field for investigation, but nevertheless could reveal some fundamental biological facts. Initially the suggestion needs to be confirmed by accurate measurement of plants grown under controlled conditions.

Does Chemically Grown Food Contain More Water?

This question should be cleared up by extension of the type of analysis previously carried out on Haughley crops.

Release of Plant Nutrients in the Soil: to What Extent is this Governed by the Crop Grown?

Pointers from the Haughley Experiment indicated that seasonal changes in soil under grass were less than with an arable crop. (See graphs, pp. 227, 228.) This line of investigation could be followed using the techniques as previously applied.

The above are some suggestions for the furtherance of useful research work into a future Haughley Experiment based on the closed-cycle principles.

It must be obvious that sufficient has been said to indicate that this was an ecological experiment and not one to indicate how to induce a greater bulk of crop. Nevertheless, the fundamental information that such a study has revealed encourages one to hope for its revival since the information obtained must benefit our overall knowledge of the biological implications of organic farming and horticulture.

We feel that a future ecological investigator could well benefit and obtain some inspiration from reading this account of what undoubtedly was a pioneer venture.

8

POSTSCRIPT
Soil, Food and Health[1]

The following statement has been sponsored by medical and dental members of the council of the Soil Association and has been signed by more than four hundred members of the medical and dental professions. The signatories include general practitioners, consultants, medical officers of health, and university professors.

THE TEXT

In 1939 the Local Medical and Panel Committee of the County of Cheshire, representing some six hundred general practitioners, issued a 'Medical Testament'. This testament, in our considered opinion, ranks as an historic document. It declared forthrightly that, whilst good progress had been made towards the second objective of the National Health Insurance Act of 1911, in so far as postponement of death could be regarded as evidence of the cure of sickness, no such progress could be observed in the attainment of the first objective—presumably of at least equal importance—the prevention of sickness. It affirmed that the basis of such prevention was essentially nutritional, its signatories stating that their daily work brought them repeatedly to the same point: 'This illness results from a lifetime of wrong nutrition.'

Medical Significance of the Organic Movement

The 'Medical Testament' was an inspiration to many forward-looking people, in particular Sir Albert Howard and Lady Eve Balfour, pioneers of what has come to be known as the Organic Movement. It confirmed and enlarged their own experiences in the field of agricul-

[1] Reprinted from *The Lancet*, 19 January, 1957.

ture and constituted a notable contribution to the thesis for which the movement stands and for which considerable support has been gained in many different parts of the world in the last ten years.

This thesis is, basically, the concept of organic wholeness applied to the vital relationships between soil, plant, animal and man, with special reference to health. It envisages health as a positive process which is profoundly influenced by the interdependence of all life, primarily through the medium of nutrition. It postulates, therefore, that the key science is the somewhat neglected one of ecology—the study of the relationships between living creatures and between such creatures and their environment.

In all these respects the Organic Movement represents a marked departure from the present trend towards the increasing fragmentation of science, the piecemeal application of science through technology as expediency dictates, and the unceasing struggle to deal with the ever-changing manifestations of individual diseases, whether in plants, animals, or humans. On the other hand, its tenets are in keeping with the definition of health propounded by the World Health Organization—namely, 'a complete physical, mental and social well-being, and not merely the absence of disease or infirmity'.[1]

Moreover, the concept of nutrition as a flow of nutrient materials from the soil through plants and animals to humans has been strongly reinforced by recent researches in the field of human metabolism. These, by the use of isotopes to trace the passage of food constituents through the body, have demonstrated that all human tissues undergo constant renewal.

It follows that the integrity and biological quality of the nutrient materials used for this renewal are of first-rate importance to health; and that the extent to which they are influenced by agricultural practices becomes a matter of great practical significance to the medical and dental professions. Yet, so far as we are aware, the only scientific studies hitherto attempted in this field are those being conducted at the Haughley Research Farms in Suffolk. These are maintained by the Soil Association as a purely voluntary effort to enlarge our stock of knowledge, independently of state aid or commercial patronage.

The Haughley Experiment is designed to investigate and compare, on a farm scale—and therefore on a realistic basis—the cumulative effects of three contrasted types of soil management on: (a) the fertility of the land, (b) the nutritional quality of its produce, and (c) the health

[1] From the first of the Nine Principles of the World Health Organization.

and productivity of farm animals fed continuously and exclusively on that produce.

[*Here followed a brief description of the layout of the Experiment and its findings up to that time.*]

Declaration

We, the undersigned medical and dental practitioners, desire to re-affirm the conviction expressed by the Cheshire Panel Committee regarding the intimate connection between nutrition and health.

We deplore the fact that, in the seventeen years that have elapsed since the publication of the 'Medical Testament', so little has been done to implement the avowed intention of the National Health Service Act, 1946, in regard to the prevention of disease, the more especially in view of the steadily increasing incidence of degenerative diseases, which contrasts markedly with the mastery achieved over infectious and inflammatory diseases.

We would draw attention to the fact that, during this period, we and a growing number of professional colleagues, together with many veterinarians, farmers, horticulturists, and others, concerned with the health of livestock and crops, have become increasingly aware of the fundamental importance of the relationships between soil, plants, animals and man.

We believe that the task of preventing disease is best envisaged as the enhancement of positive health and that, in this task, nutritional research should play a leading part.

We believe, further, that any such research which fails to take into consideration the soil on which the food is grown, and the farming practices involved, must be accounted incomplete. This belief is in no way modified by the fact that much food is subjected to chemical and mechanical treatment after it leaves the land, flour being an outstanding example. Such treatment also is of deep concern to us, though it cannot be discussed here. We would, however, observe that it may well depress still further any reduction in food values that may be attributable to current agricultural practices.

We would urge, therefore, that any research designed to discover which methods of soil management result in the highest food values should receive our professional support, and that such research should be both comprehensive in scope and of long term—extending over several decades rather than years.

We note that apparently the only such research at present in progress is that being conducted at the Haughley Research Farms already mentioned, and consider that it would be a tragedy if this one experiment with its prepared material, were to be brought to a premature end—as seems only too possible—for lack of public interest and support. It has now reached a stage at which its further extension, especially on the scientific side, should yield an increasing dividend of knowledge; but the enthusiasm of the Organic Movement, which has already accomplished so much, cannot be expected to undertake unaided the heavy financial commitment involved.

In support of these contentions, we would cite Sir John Charles who, in his 1952 annual report as Chief Medical Officer, Ministry of Health,[1] referred to the uncharted fields, exploration of which would surely reduce the burden of the curative services, and who called for research into the causation of the so-called chronic diseases. Sir John referred also to the activities of the Medical Research Council, indicating that the fields of research lying open to the investigators would not be narrowly limited.

We would cite also the Chief Medical Officer for the Department of Health for Scotland who, in his 1955 annual report,[2] states that 'at later ages, the increasing threats of coronary thrombosis and cardiovascular disease, and of cancer of the lung, especially in males, maintained their position as major health problems in which no clear lines of medical attack have yet emerged.'

In our view, the recent suggestion of the importance of the essential fatty acids to the circulatory and nervous systems and to the carcinoma barrier points to the kind of knowledge which is becoming essential to us as practitioners.

We, therefore, call upon those responsible for the finance, initiation, and guidance of medical research to consider the urgent need for ecological research, linking the basic approach to human nutrition with the origins of human food and, in particular, to assist the one experiment already in progress in this field.

[1] Annual Report of the Ministry of Health, II, 1952, p. 17, HMSO, 1953.

[2] Annual Report of the Chief Medical Officer, Department of Health for Scotland, 1955, HMSO, 1956.

B. P. Allinson, MRCS
G. W. Anderson, MS
J. C. Anderson, OBE, TD, MB, FRCS
Mary Anderson, MB, DPH
Patrick H. R. Anderson, BSc(Agr), MD, DObst, DPH
Desmond J. Atherton, MB, MRCS
A. S. Baker, MB
J. B. Bamford, MRCS, DA
Mary Barton, MB
A. Priestley Bates, MB, DObst
M. Beddow Bayly, MRCS
R. L. Bell, MB
J. Barry Bennett, MB
L. Berlanny, MB
C. Binney, MB
R. A. Binning, FFA, RCS
Oliver H. Blacklay, MD, FRCS
John Boardman, MB
N. A. Boswell, MB, MRCS
G. E. Breen, MD, DPH
C. M. Brooks, LMSSA
D. H. Broughton, MB, MRCS
Richard Brown, MB, DPH
H. J. M. Browne, MRCS
J. L. Burn, MD
A. V. Campbell, MD, DPH
E. L. Campbell, MB
J. D. C. Campbell, LRCPE
O. M. Capper-Johnson, MA, MB, MRCP
E. R. Carter, MB
A. G. Chamberlain, MRCS
H. J. Churchill, MRCS
H. M. Clark, MB
S. A. Clark, MA, MD
R. T. V. Clarke, MB, MRCS
H. H. Corrigall, TD, MB, FRCP
E. E. Critchley, MB
J. S. Cruickshank, MB
A. B. Cunning, MB
T. M. Curran, MD, MRCP
B. Dale, MRCS
G. Darke, MB
R. Baring Davidson, MB, LRCPE, DPH
George Day, MB
Guy Daynes, MRCS, DCH
F. A. Dick, MB
U. M. Dick, MB
Harold Dodd, ChM, FRCS
M. T. J. d'Offay, MB, FRCS, RAMC
A. J. H. Donnell, BA, MB
A. J. Drew, MB, FRCS
E. A. Eason, MB
H. Tudor Edmunds, MB, MRCS
H. W. Featherstone, OBE, TD, MD, FFARCS, JP
J. H. Ferguson, MB
A. E. Fiddian, MRCS
W. P. Foster, MB, MRCS, DCH
J. Boyd Fulton, MB
E. B. Garrett, MB, BSc, MRCS
T. M. Gibson, MB
T. Maitland Gibson, MB
P. W. W. Gifford, MB, MRCS
P. Gilbert, MB, MRCS
D. H. C. Given, MD, DPH
E. A. Green, MA, BM
K. McG. Greer, MB
H. W. Hales, MD
A. Hargreaves, MRCS
D. H. C. Harland, MB, FRCS
B. B. Harrison, MB, DMRD
E. S. Hawkes, MB
L. C. Hayes, MB, DPH, RAMC
J. L. Hine, MA, MB
G. R. Holtby, MD, DPH, DIH
J. P. Hope, MB
L. Hornung, MD, MRCP, DCH
J. H. Horsley, MB
P. M. Horton, MRCS, DPM
Grenville Hoyle, MB
C. A. Hutt, MRCS
J. H. Inskip, MA, MB
S. A. Jenkins, MB, FRCS
B. C. Jennings, MRCS
B. Johnson, FFARCS, MRCS
B. McDougall Johnson, MB
D. M. M. Jones, MB, MRCS, DPH
H. W. Francis Jones, MB, MRCP
M. T. Islwyn Jones, MD, DPH
R. Francis Jones, MB
H. W. Jordan, MB, MRCS
T. B. Kenderdine, MB, MRCS, DA
John Kerr, MB
Ronald Kerr, MRCS, DOMS
H. P. Kilsby, LRCPI
L. E. D. Knights, FRCS
W. H. L. La Frenais, BSc, LRCPE
Conrad Latto, MB, FRCS
D. A. Latto, MB, MRCOG
Gordon Latto, MB
Monica Latto, LRCPE, DObst
Geo. Laurence, FRCS

Rupert Laverty, MB
K. A. W. Law, MBE, MB
L. B. Lawrence, MB, DPH
E. K. Ledermann, MD, LRCPE
O. Leeser, MD, PhD
H. L. Lentin, MB
R. I. Lindfield, MRCS
Hugh L. Mackintosh, MB
John Maddison, MD, DPH
O. E. Manasse, MD, LRCPE
S. B. Mathews, MB
N. J. McAllister, LRCPI
J. E. R. McDonagh, FRCS
Winifred McIlwrick, MB
H. Meacock, MB
T. R. C. Melrose, FRCS
E. F. Meyer, MD, LRCPE
A. H. Michael-Phillips, MB, MRCS
M. L. Millard, MB
C. E. Miller, LRCPE
Ivor H. Mills, BA, MB, PhD, BSc, MRCP
A. P. Milner, BA, MB
A. R. Milson, MB
F. E. Milson, MB
Olive B. Milson, MB
E. Moore, MB
W. G. Moore, MB
J. Morgan, MB
T. B. Moriarty, LRCPI
Donald S. Morris, MRCS
G. E. Mullins, MRCS
P. Murphy, MB
O. M. Naylor, MB
Ivan D. M. Nelson, MB, DPH
G. H. Newell, MB, DTM & H
H. A. Nicholls, MRCS
A. Noll, MD, LRCPE
C. Noon, OBE, FRCS
P. M. Oxley, MRCS
M. G. Paine, MB
A. C. Palmer, FRCS, FRCOG
C. G. Pantin, MD
Dorothy Pantin, MB, MRCS
Miles Parkes, MB
P. W. Parkes, MRCS
J. Parrish, MB
J. A. Parrish, MB
S. Chalmers Parry, MRCS
W. D. Paterson, MB, FRCS
R. W. Payne, FRCS, DOMS
Innes H. Pearse, MD, LMSSA
J. H. Penrose, BA, MB, FRCS
N. C. Penrose, OBE, MB
J. Pinching, BM, MRCP
C. V. Pink, MRCS
D. S. Piper, MB
L. Duncan Porteous, MRCS
S. G. Ransom, MRCS, DA
M. L. Rawlins, MB
E. Harford Rees, FRCSE, MRCOG
H. G. H. Richards, MD
G. C. D. Roberts, FRCS
H. D. Robertson, MRCS
R. Roderick, MRCP
Charles D. Ross, MD
T. H. Sanderson-Wells, MBE, MD, FRCS
E. Saunsbury, MB
A. Savill, MD, FRCP
J. W. Scharff, MD, DPH, DTM & H
A. D. Scott, MB
D. A. Sharpe, FRCS, MRCOG
C. H. Shaw, MD, DPH, DPA
L. G. Shearer, MB
J. D. Shed, MB
A. L. Slater, MB, BSc
John Sleigh, MB, DPH
J. A. Small, MA, MB
J. H. W. Smith, MB, FFA, RCS
W. D. A. Smith, OBE, MD
W. R. Spacek, FFA, RCS
P. Q. M. Spaight, MRCS
T. T. Stamm, MB, FRCS
E. E. Stephens, MD
T. R. Stevens, FRCS
H. Stranz, MD
B. R. Sworn, MB, FRCS
Anne M. Timmis, MB
H. Tomlinson, MB, MRCS
M. J. Townsley, FRCS
J. B. Tracey, MB
Sybil Tremellen, MRCS
D. C. L. Vey, MB
K. O. A. Vickery, MD, MRCS, DPH
Cecil Wakeley, BT, KBE, CB, DSc, FRCS
J. K. Watkin, MB, DPH
A. T. Westlake, MB, MRCS
E. K. Westlake, MA, MD, MRCP
J. B. Wheeler, MB, MRCS
A. P. Whitfield, MB
A. D. Willis, MB, MRCS
E. G. Wilson, MB
G. I. Wilson, MB, FRCS
I. Wilson, MD, FRCP, DPH
J. Greenwood Wilson, MD, FRCP, DPH

R. Allan Wilson, MD
W. Winch, MA, MRCS
John Winning, LRCPE
C. J. J. Winter, MB
A. Woolley, MBE, MB, MRCS
Eric Wordley, MD, FRCP
E. W. Wright, MB, DPH
C. H. Wrigley, MD
W. W. Yellowlees, MB

Overseas

G. E. V. Conran, BSc, MB
C. R. Copland, MB
William A. Daniels, MD
J. B. David, MB, FRCS, DLO
J. Deakin, MB, MRCP
A. W. N. Druitt, MRCS
Ray Evers, MD, MS
L. Glaser
Glauert-Hoffmann
Gondolatsch
T. Dudley Hagger, MD
C. S. Hamner, MD
Hofmann
M. Percy Jackson, MB, MRCS
C. Johnson, MD
Mary Johnstone, LRCPE, DPM
Kumpf
R. H. Little, MD
Tom Melton, MD
Francis M. Pottenger, MD
Miles H. Robinson, MD
Rouenhoff
D. P. Rowe, MB
Hanns Peter Rusch
Vernon Stabler, MD
H. H. Stott, MB
Stumpfig
John Wade, Jun., MD
Graeme Williams, BSc, MB
Tom Williams, MD

Dental Surgeons

R. A. Abbey, LDS, DDO
Joyce H. Aitken, LDS
J. B. Andrew, BDS, LDS
G. R. Barnes, BDS, LDS
P. R. Barton, BDS, LDS
W. J. Bate, LDS
P. V. Bax
Robert T. Bax, LDS
D. Beacher, BDS
M. G. Berry, LDS
E. C. Betts, LDS
A. A. Blake, LDS
J. P. Blunt, LDS
M. Harvey Booth, BDS, LDS
L. H. Bradbeer, LDS
J. A. Brammer, BDS, LDS
B. R. Burn, LDS
C. M. Burleigh, LDS
A. Courtney Campbell, BDS
C. Douglas Campbell, LRCPE, LDS
J. Campbell, PhD, FDS, DDO
Violet Campbell, LDS, RFPS
W. G. Campbell, LDS
N. S. Cardell, LDS
E. Brodie Carpenter, LDS
H. G. Carr, LDS
G. K. Catchpole, LDS
John Chalmers, MBE, FDS, JP
P. A. Chandler, LDS
D. J. Chant, BDS, LDS
Donald A. Clark, FDS
M. H. Clutterbuck, LDS
T. C. Collins, LDS
Bernard Cooke, LDS, FRIPH
Ernest Cooper, LDS
Jas. B. Coventry, LDS
R. J. Cowling, LDS
H. L. Davies, LDS
S. W. A. Davis, LDS
K. McL. Dorning, LDS
Frank H. Edey, LDS
Peter A. Edge, LDS
G. H. Ellingham, LDS
P. Ellis, LDS
L. W. Elmer, LDS
R. W. Emerson, LDS
N. S. Farnes, LDS
Alexander Ferguson, LDS
E. C. Fox, MD, LDS

Dental Surgeons Cont.
J. A. Gale, LDS, DPA
David Gent, LDS
G. D. Gibling, LDS
Philip Glickman, LDS
Rodney S. Glickman, LDS
E. A. Glover, BDS, LDS
P. W. Goodall, LDS
H. E. Gray, LDS
Norman Gray, FDS, HDD
H. G. Griffith, BDS, LDS
W. B. Hales, FDS
C. E. M. Hallett, MDS, FDS, HDD, DOrth, RCS
R. K. Hamilton, LDS
E. E. M. Hansom, LDS
J. L. Hardwick, MDS, FDS
W. F. Hinds, LDS
M. I. Hoggins, LDS
Keith Holman, LDS
J. G. Hopkinson, BDS
W. A. W. Howe, LDS
R. E. Hyman, LDS
J. V. Inglis, LDS
G. Jeffery, LDS
Thomas John, LDS
F. K. Johnson, LDS
Geo. Johnson, LDS
E. Leslie Jones, LDS
H. T. Jones, LDS
W. E. Jones, LDS
H. Fawcett King, LDS
M. I. Lamb, LDS
H. Matheson Latto, BDS
H. M. Laurence, LDS
F. F. Leek, LDS
L. C. L. Lees, LDS
W. H. Liebow, LDS
Edward Line, BDS
J. J. Long, LDS
Oswald A. Long, LDS
R. W. Loval, DDS, FDS, HDD
R. Lovewell, LDS
David Lumsden, LRCPE, LDS
D. MacLaren, LDS
Norman R. Marsden, LDS
L. E. Marshall, LDS
J. W. Martin, LDS
K. McAllister, LDS
A. W. McCandlish, LDS
Janet A. McCann, LDS
Alastair McCully, LDS
D. M. McDiarmid, LDS
S. P. Meacock, MRCS, LDS
G. C. Mence, LDS
J. M. H. Miller, LDS
Reginald Miller, LDS
A. C. Mockridge, LDS
B. R. Moore, LDS
T. A. Morris, BDS, LDS
A. A. Morse, LDS
C. Nest, LDS
P. A. Newman, LDS
J. Nicolls, LDS
J. D. O'Brien, LDS
A. O'Connor, BDS
E. P. Pamment, LDS
H. Phillips
P. L. Pickett, LDS
R. W. Pinniger, LDS
A. Pirrie, LDS
Morris Powell, LDS
J. F. Powell-Cullingford, LDS
P. J. Pretty, LDS
N. W. Ramage, LDS
A. W. Reynolds, LDS
G. D. Richards, LDS
W. K. Rimmer, LDS
D. B. Robertson, LDS
D. Robertson-Ritchie, HDD, LDS
A. E. Robinson, FDS
A. B. Rodger, LDS
K. W. Rose, LDS
Allan Ross, MRCS, BDS
E. Runeckles, LDS
Henry P. Sager, LDS
Edward Samson, FDS
D. M. Sanderson, LDS
J. A. Hoggarth Savage, LDS
E. M. S. Scanlan, LDS
Albert Scruton, LDS
A. F. D. Shapland, LDS
E. M. Smallwood, LDS
J. A. Clayton Smith, LDS
T. Morrish Smith, LDS
D. R. Stevens, BDS
W. H. Taylor, LDS
C. F. Teall, LDS
G. E. M. Teall, LDS
W. Thom, LDS
O. S. B. Thomas, LDS
J. C. Timmis, LDS
C. Grenville T. Tregarthen, FRHS, MDS, FDS
Everard Turner, LDS
J. W. Turner, LDS
R. Waldsax, DMD
L. V. Wardle, BDS, LDS

Dental Surgeons Cont.
H. G. Watkin, FDS
Elizabeth Watson, LDS
D. B. Waugh, LDS
Elizabeth M. Webster, LDS, DDO
G. Campbell Webster, LDS
D. B. Wells, LDS
Donald Westoby, LDS
P. H. Williams, LDS
J. F. Wheatcroft, LDS
H. Woodhead
Frank M. Wood, LDS
H. A. Woodward, LDS
W. E. Woolcott
W. H. Wotton, LDS
T. J. Wright, MDS, LDS
L. A. Wyborn, LDS
G. M. Yardley, FDS, HDD
R. Miller Yardley, LDS

APPENDICES, BIBLIOGRAPHY AND RECOMMENDED READING, GLOSSARY and INDEX

Appendices

APPENDIX 1

Administration

THE SOIL ASSOCIATION
(1947–71)

President and Chairman of Council: The Earl of Bradford (1951–70)

Subsidiary Operating Company
Haughley Research Farms Ltd
Chairman: Sir Cecil Oakes, CBE (1948–59)
Chairman: J. P. Marland, JP (1959–69)

Director of Research: R. F. Milton, BSc, PhD, FRIC, FI Biol (1952–68)
Assistant Scientific Officer (Part-time): J. F. Ward, MA, Dip Hort

RESEARCH ADVISORY COMMITTEE
(1967–70)

Chairman: Dr. K. Mellanby

Members: Professor R. L. Robb; Dr. R. F. Milton; Dr. Innes Pearse; Lady Eve Balfour; J. P. Marland; Dr. R. L. Jefferies; Dr. G. S. Hartley; Mrs. M. A. S. Bates; Lt.-Col. Sir Robert Milnes Coates, Bt.; P. R. Peecock.

Secretary and Farm and Research Administrator:
D. D. Campbell, OBE, MC, BSc (1965–70)

Scientific and Agricultural Consultants

The deliberations of the committee, appointed by the Soil Association to lay down the framework for the second stage of the Experiment, were conducted under the chairmanship of the late George Scott-

Williamson, MD, research biologist.[1] Two other members with ecological research experience were the late Dr. Rayner[2] and her husband, Professor Neilson-Jones.[3] Three well-known farmers were brought in as consultants: Mr. Friend Sykes from Wiltshire for the Organic section; the late Mr. William Alexander from Kent for the Mixed section, and the late Mr. E. P. Chamberlain from Oxfordshire, for the Stockless section. Other people consulted included the late Sir George Stapleton, and the late Sir Robert McCarrison.

[1] Founder of the Pioneer Health Centre, Peckham. See I. H. Pearse and L. H. Crocker, *The Peckham Experiment*, Allen & Unwin, London, 1943.

[2] Botanist and mycologist. Author of *Mycorrhiza: an account of non-pathogenic infection by fungi in vascular plants and bryophytes*, Cambridge University Press, 1927.

[3] Professor of Botany, Bedford College, University of London, 1920–48.

APPENDIX 2

Additional Farm Data

CROPPING SYSTEMS FOR MOST OF THE PERIOD

(all figures are per acre)

Year in rotation Crop	ORGANIC SECTION *Treatment*	MIXED SECTION *Treatment*
1. Winter Wheat Undersown for autumn graze and plough back. Sub-soiled after harvest	nil	3–4 cwt. NPK at sowing 1–2 cwt. N in spring
2. Kale and Maize	10–12 tons composted FYM	12–15 tons FYM 9–10 cwt. NPK at sowing
3. Spring Barley	nil	3 cwt. NPK at sowing
4. Winter Beans and Spring Peas	nil	3–4 cwt. PK at sowing
5. Winter Oats Undersown[1] for autumn graze and plough back. Sub-soiled after harvest	nil	3 cwt. NPK at sowing
6. Silage Mixture (oats and peas) Undersown with four-year Ley mixture	10–12 tons composted FYM	12–15 tons FYM 3–4 cwt. NPK at sowing
7. 1st year of Ley	Strip-grazed by cows	Strip-grazed by cows 3 cwt. NPK Possibly N in summer
8. 2nd year of Ley	Cut for hay Grazed by cows and sheep	Cut for hay Grazed by cows and sheep 3 cwt. NPK in autumn 2 cwt. N in spring

[1] See p. 300.

9. 3rd year of Ley	Folded with poultry Grazed by cows and sheep	Folded with poultry Grazed by cows and sheep Slag in winter
10. 4th year of Ley	Grazed by cows and sheep Broken up in August	Grazed by cows and sheep 2 cwt. N in spring if required Broken up in August

STOCKLESS SECTION

Year in rotation	*Crop*	*Treatment*
1.	Wheat	3 cwt. slag on clover stalks before ploughing in 3 cwt. PK at sowing 2–3 cwt. N in spring. Field subsoiled after harvest
2.	Sugar-Beet	10–12 cwt. NPK at sowing (Sprayed with Metasystox for aphid if necessary) Beet tops ploughed in
3.	Barley	4 cwt. NPK at sowing
4.	Barley undersown with clover 1 acre Beans	3 cwt. slag on barley straw silorated before ploughing in 5 cwt. NPK at sowing 3 cwt. slag on barley straw silorated before ploughing in 4 cwt. PK at sowing
5.	{ Clover 1 acre of half fallow, i.e. cultivated till mid-summer then mustard sown and ploughed in	Barley straw silorated and left as mulch Clover topped and dropped Crop of clover seed taken, stalks left

N = Nitrogen
P = Phosphate } as chemical fertilizer.
K = Potash

Silorator = Cutter-blower.

For details of fertilizer application, see table on p. 330.

FYM = Farm Yard Manure.

One field outside rotation Lucerne (Alf-Alfa)	ORGANIC SECTION *Treatment* Composted FYM once in the six years if possible	MIXED SECTION *Treatment* FYM once in six years if possible 3 cwt. NPK every autumn and/or spring

Normal annual cropping of lucerne ley on both sections:
1st cut, silage
2nd cut, hay
3rd cut, hay or grazing.

After six years the field returns to rotation and another field drops out for lucerne.

It will be noted from the table on p. 297 that rotations and general ley management are identical in both stockbearing sections. The varieties of the crops are also the same, with one exception. On the Organic section a number of deep-rooting herbs are included in the ley mixtures, in accordance with the theory that when no mineral fertilizers are applied, natural soil minerals (including trace elements) must be kept in circulation by biological means, of which the use of such special 'wild' plants is one.

CROP VARIETIES IN USE

Crop	*Variety*	*Year of purchase of original seed*	*Generation (1961)*		
			Organic	*Mixed*	*Stockless*
Wheat	Atle	1948	14th	14th	14th
Beans	Tic Beans	1948	14th	14th	14th
Peas	Dun Peas	1948	14th	14th	*not grown*
Oats	S.147[1]	1953	9th	9th	*not grown*
Barley	Rika[2]	1956	6th	6th	6th
Maize	½ King V, ½ White Horse-tooth	every year	—	—	*not grown*
Kale	Thousand-head	every year	—	—	*not grown*

[1] Before 1953, Star.

[2] Before 1956, Spratt Archer.

4-YEAR LEY MIXTURES
(per acre)

ORGANIC[1]	*lb.*
Kentish Indigenous Ryegrass	5
Timothy S.48	3
Timothy S.51	2
Timothy (Scotch)	2
Cocksfoot, Leafy S.37	2
Cocksfoot, Leafy S.143	3
Rough Stalked Meadow Grass	½
Meadow Fescue	4
Tall Fescue	2
Common Milled Sainfoin	3
American Sweet Clover (*Melilotus alba*)	2
Late-Flowering Red Clover	1
Alsike Clover	2
Kentish Indigenous Wild White Clover	¼
Kidney Vetch	1
Yarrow	¼
Burnet	3
Chicory	1
Sheep's Parsley	1
Total per acre	38

MIXED	*lb.*
S.24 Perennial Ryegrass	4
Danish Perennial Ryegrass	3
S.215 Meadow Fescue	3
Canadian Meadow Fescue	6
S.51 Timothy	3
Canadian Timothy	2
Smooth Stalked Meadow Grass	1½
Late-Flowering Red Clover	1½
Kersey White Clover	2
Total per acre	26

S. denotes Aberystwyth-bred strain.

[1] This mixture has been modified for the 1962 sowing.

UNDERSOWN 1-YEAR LEY FOR PLOUGH-BACK
(see table on p. 297)

ORGANIC *and* MIXED	*lb.*
Irish Perennial Ryegrass	12
Broad-Leaf Red Clover	2
Kersey White Clover	½
Total per acre	14½

STOCKLESS	
Until 1948	Yellow Trefoil
From 1949	Broad Red Clover
(*From* 1962 Late	Flowering Red Clover)

Although Atle was originally introduced as a spring wheat, it is the custom at Haughley to sow it in the autumn. It behaves very satisfactorily when treated as a winter wheat. The fact that it can be sown in either spring or autumn has a great advantage in that, in a very wet season, should autumn sowing fail, there is a second chance in the spring. This is an important consideration where home-grown seed has to be used, though in practice that expedient has had to be adopted only once.

Livestock Populations

The number of stock as at the end of 1961, was as follows:

	Organic	*Mixed*
GUERNSEY CATTLE		
Cows, including heifers due to calve	15	16
Young Stock	8	13
Bulls	1	1
CLUN FOREST SHEEP		
Breeding Ewes	13	13
Rams	1	1
LIGHT SUSSEX POULTRY	250	300

Note: The reason for the small number of heifer calves on the Organic section, at the end of 1961, was an unusually long run of bull calves. Normally the same or similar numbers of cattle are kept on both sections.

Sires

The bulls, purchased as calves and reared, are shown above as one in each section; but the method of management is for each section to carry both bulls in rotation, since both are used on both herds in order to reduce differences arising from genetic factors.

Ram lambs will be purchased in alternate years, and change flocks in the second year.

Cocks are bought each year from W. Golden of Leire, whose system of management is followed, those allocated to the breeding pens of one flock being full brothers of those allocated to the other.

Numbers of Breeding Stock per Section

CATTLE fluctuate a little, but somewhere between 25–30 per section (all ages) seems to be the capacity.

EWES will be increased, probably up to 20–25.

HENS. Up to the present it has been possible to maintain rather more on the Mixed section than on the Organic (see table on p. 302).

Total Numbers

In calculating total carrying capacity, it must be remembered that, from April to August or September inclusive, poultry numbers are doubled, since the older birds are not culled until the pullets are on the point of lay.

Similarly, from March to July, sheep numbers may be trebled in a good lambing season.

LIVESTOCK POPULATIONS FOR THE LAST SEVEN YEARS OF THE EXPERIMENT

Date	*Cattle*		*Sheep*		*Poultry*	
	Organic	*Mixed*	*Organic*	*Mixed*	*Organic*	*Mixed*
1.4.63	26	28	14[1]	17[1]	246	295
15.9.64	33	30	21	21	257	305
31.12.65	26	30	18	23	220	268
31.12.66	26	30	25	26	304	326
31.12.67	24	28	26	31	242	354
31.12.68	30	37	26	36	498[2]	531[2]
31.12.69	33	30	26	35	236	343

[1] Excluding 1963 births. [2] Before the annual cull.

Reminder: All livestock were maintained all the year round exclusively on the produce of their own section of approximately 75 acres.

Method of Establishing the Herds and Flocks

Twenty-two foundation Guernseys were bought in 1948 and 1949, two as in calf heifers, eight as bulling heifers, and twelve as yearlings. They all came from Cornwall, and were purchased on the Directors' behalf in pairs of animals, as closely related as possible—never less than half-sisters—and out of dams of similar line-breeding and production records. On arrival, Mr. Friend Sykes of Chantry for the Organic section, and Mr. William Alexander of Kent for the Mixed section, picked, in turns, one of each pair of sisters for the section they represented. All the animals in the present two herds originated from those few original heifers, and the same bulls have been used in both herds in each successive generation, so genetically the two herds have tended to get more closely related rather than less.

The foundation Light Sussex hens were bought in 1952, fifty per section, from the same batch of chicks. They have been bred up to their present numbers using Golden cockerels each year, so that now, in their ninth generation, they are nearly pure Golden Strain.

Eight of the foundation Clun Forest ewes were purchased as shearlings in 1960, from the Institute of Animal Physiology at Babraham; and eighteen in 1961, eight two- and three-year-olds from the Rushbrook flock, and ten old cull ewes from Tibbets of Huntingdon. The ram lambs came from Babraham. The first eight were carried on the Organic section during 1960 but in 1961 each of these three small consignments of sheep were divided equally between the two sections. All foundation sheep were paid for by an Australian supporter.

All the livestock, as already stated, are fed exclusively on the produce grown on the section to which they belong, but small quantities of food supplements are imported. The rations of both flocks of poultry include rather less than 5 per cent of fishmeal. On the Mixed section *only*, 2 per cent of mineral mixture and 3 per cent of fishmeal is added to the grain ration given to cattle, and all cattle and sheep on that section have access to mineral licks. On the Organic section dried seaweed is substituted for the conventional minerals. Both herds of cattle have access to rock salt. For details of rations, see pp. 305–308.

Farmyard Manure and Compost

On both sections the milking cows, and young stock up to the age of yearlings, are housed in yards during the winter, though the cows go out during the day. Bulling and in-calf heifers usually winter out. All the bedding straw, of course, comes only from their own section, and the resulting farmyard manure is returned only to its own section. The yards in which straw and manure have accumulated during the winter, are cleaned out in the summer. On the Mixed section the material is occasionally put straight on to a field, but usually it is first assembled in 'hill' for several weeks. On the Organic section, it is made into rough compost—that is to say, a small proportion of green weeds and soil are added when the 'hill' is made; this is built up in sections, each measuring roughly 12 ft. × 4 ft. × 4 ft. and in each of which two vertical ventilation holes are made. The 'hill' is turned twice if possible, but always once.

This compost, and the rotted FYM on the Mixed section, is applied twice in the arable part of the rotation. In recent years, it has gone to the maize and kale in the second year, and again in the sixth year for the arable silage, which also forms the nurse crop for the four-year ley. In earlier years it was applied to the beans and roots, and to the second-year ley.

Weed and Pest Control

On the Mixed and Stockless sections hormone weed-killers are used when the need arises, as also are insecticides. On the Organic section no chemical sprays are used. Insect pests have so far been no problem on this section, except for turnip flea beetle, which is dealt with by avoiding mid-season sowing of brassicas. In the case of weeds, the aim on this section is control but not elimination. So long as they are not permitted to compete with the crop to the point of injury, some weeds are considered to be beneficial rather than harmful. The methods of control used are good cultivations, mechanical hoeing, and encouragement of the crop so that it smothers the weeds.

Seed Dressings

On the Mixed and Stockless sections, mercuric seed dressings are used on all cereal and pulse crops, destined for ripening. No seed dressings have been used on the Organic section, except in certain years on wheat, starting with the first three generations. Then, after six generations without dressings, the crop was affected by bunt in a particularly bad season for fungus attack. Since then, part of the wheat crop has had the seed dressed, and part has been heat treated, in order to preserve the continuity of an undressed strain. Chemical dressings will be dropped again as soon as possible.

Ley Management

Management of the leys on both sections consists of a light grazing in the early autumn of the year it is sown, after the silage crop comes off. It is then rested during the first winter, and strip-grazed by the cows during the following spring and summer. The renewed growth of herbage between grazings is allowed to reach 6 in. high, and as many as six strip grazings may take place during the season. At least once during the first season the ley is 'topped', and it is harrowed at the beginning of most rest periods, sometimes with chain harrows, but always at least once with pitchpole harrows, using the cutting blades.

During its second winter, after a light autumn grazing by sheep, the ley is again winter rested, and hay is taken from it in its second year, the crop being dried on tripods. After the field has been cleared the poultry are moved on to it. These are housed in free-range houses on skids, inside an electric fence enclosing about two acres of the ley at a time. The houses are moved forward each week, and when they have covered the two acres, the whole enclosure moves on. In this way the hens continually graze, and scratch, and manure the entire ley in the course of twelve months, and other stock can be grazed at the same time, either in front of or behind the hens according to the state of the herbage. During this third year, winter as well as summer grazing takes place, and is fairly severe. After recovery in the spring, the ley, now in its fourth year, is grazed (with rests as required) by cattle and sheep until it is broken up for wheat in the late summer. Whenever possible, this is done by first chopping up the turf with a rotavator, and then using a chisel plough. (The gift of an Australian supporter.) The mouldboard plough, which inverts the soil, is avoided as far as possible on land to be used for autumn sowing.

Manuring

This system of ley management, and the two applications of compost during the 9–10-year rotation, is all the manuring the Organic land receives. The Mixed section receives, in addition to its own FYM, fertilizers every year in the rotation; quantities and NPK ratio varying according to crop and season. For details, and also for the Stockless section, see the table on p. 330, and pp. 297–8.

RATIONS

Cows

In a normal season the feeding system we adopt is as follows:

Approximately mid May to mid August, grazing only except for any cow giving 4 gallons or over; from mid August to the end of October, autumn grazing supplemented by maize cut and fed green; November to March inclusive, full winter rations; April and early May, half winter rations.

Winter rations consist of kale and/or silage, hay, oat and barley straw, and home grown, home ground, concentrates. The composition of the meal varies from year to year, depending on available supplies.

COMPOSITION OF HOMEGROWN CONCENTRATES

(as percentages)

	WINTER 1961–2		WINTER 1962–3	
	Organic	*Mixed*	*Organic*	*Mixed*
Oats	21	30	30	30
Barley	54	30	55	30
Beans or Peas	25	25	—	25
Wheat	—	10	15	10
Fishmeal	—	3	—	3
Minerals	—	2	—	2

In 1961–2 oats were reduced and wheat omitted on the Organic section, because these two items of food were in short supply. In 1962–3 the shortages were wheat on the Mixed section, and pulses on the Organic section, and the rations were adjusted accordingly.

All the year round, mineral licks are made available for the Mixed-section cattle and seaweed meal for the Organic. Both are given ad-lib. In practice seaweed consumption rarely reaches 1 lb. per head per day.

DAILY ALLOWANCE OF CONCENTRATES

(lb.)

(Home-grown cereals and pulses)
WINTERS OF 1961–2 AND 1962–3

	1st gallon	*2nd gallon*	*3rd and each subsequent gallon*
Organic	Nil	2	3
Mixed	Nil	$3\frac{1}{2}$	$3\frac{1}{2}$

WINTER FODDER RATIONS AVERAGE PER HEAD PER DAY

	1961–2		*1962–3*	
	Organic	*Mixed*	*Organic*	*Mixed*
Hay	11–12 lb.	15 lb.	12 lb.	12 lb.
Silage	50 lb.	*Ad. lib.*	50 lb.	50 lb.
Straw	$5\frac{1}{2}$ lb.	$5\frac{1}{2}$ lb.	*Ad. lib.*	*Ad. lib.*

The greater amount of hay and silage fed to the Organic herd, relative to the Mixed herd, in the winter of 1962–3 is due to the complete failure of the kale crop on Organic in 1962. There was a good crop of kale on the Mixed section which supplemented the silage right through into the New Year, so that during the first half of the winter the silage fed was less, and in the second half more, than in the average figures given above.

COMPOSITION OF FEED FOR LAYING FLOCKS

(as percentages)

	SEASON 1961–2					
	SUMMER			WINTER		
	Grain		*Mash*	*Grain*		*Mash*
	O	*M*	*O & M*	*O*	*M*	*O & M*
Wheat	51	60	33	51	60	18
Oats	32	30	33	32	20	30
Barley	9	10	19	9	15	30
Peas	8		5	8	5	5
Fish Meal			10			10
Lucerne Meal						7

	SEASON 1962–3					
	SUMMER		WINTER			
	Grain	*Mash*	*Grain*			*Mash*
			Nov., Dec., Apr.		*Jan., Feb., March*	
	O & M	*O & M*	*O*	*M*	*O & M*	*O & M*
Wheat	60	33	65	60	60	18
Oats	30	33	20	20	20	30
Barley	10	24	15	15	15	30
Beans				5	5	5
Fish Meal		10				10
Lucerne Meal						7

Differences in composition of the grain feed between 'O' and 'M' were due to the shortages already mentioned, i.e. on 'O', wheat in 1961–2 and pulse in 1962–3, and on 'M', wheat in 1962–3. On neither section were any peas available in 1962–3.

In both these, and previous years, the total allowance per day per bird (grain and mash combined) was 4½ oz. to the Mixed-section flock and 4 oz. to the Organic-section flock, but the 'O' flock has access to dried seaweed.

APPROXIMATE ANALYSIS OF MINERAL MATTER IN THE SEAWEED MEAL FOR 'O' CATTLE

(as percentages)

Calcium (CaO)	1·540
Potassium (K_2O)	2·830
Iodine (I)	·170
Sodium (Na_2O)	1·930
Bromine (Br)	·680
Aluminium (Al)	·220
Phosphorous (P_2O_5)	·180
Silica	1·480
Acid Insolubles	1·000
Copper (Cu)	·002
Manganese (Mn)	·016
Iron (Fe)	·100
Sulphates (SO_4)	4·180
Chlorine (Cl)	1·720
Magnesia (MgO)	1·010

Traces of: Cobalt, Molybdenum, Boron, Vanadium, Nickel, Zinc, Chromium, Tin, Gold, Silver and many other elements

ANALYSIS OF MINERAL MIXTURE AND LICKS FOR 'M' CATTLE

(as percentages)

	Mineral powder	*Mineral licks*
Calcium	28·34	30·00
Salt (NaCl)	20·00	30·00
Phosphorus	6·75	8·00
Iron	·98	1·30
Magnesium	16·85	2·50
Sulphur	·67	—
Iodine	·22	·03
Manganese	·22	·08
Copper	1·75	·09
Cobalt	·04	·03
Zinc	·22	—
Spice and additives	·28	—

APPENDIX 3

Dr. Milton's Summary of Analyses Carried Out for Haughley, 1952–60

METHODS OF ANALYSIS

SOIL

Available Minerals. Eight-hour extraction of air-dried soil in 10 vols. of 1 per cent each citric-acetic mixture followed by eight hours mechanical shaking.
Phosphate. Spectrophotometric Molybdenum Blue method.
Potash. Flame photometry.
Nitrate. Spectrophotometric Xylenol method.
Ammonia. Alkaline Silver method.
Humus. Dichromate Oxidation for organic carbon.
Calcium. Oxalate precipitation and Permanganate titration.
Magnesium. Spectrophotometric method for Phosphate after separation as Magnesium Phosphate.
Trace Elements (Minor elements). Oxime extraction of air-dried soil for available ions, followed by acid solution of the oxidized extract.

Iron. Determined spectrophotometrically by the Thiocyanate technique.

Boron. Determined spectrophotometrically by the Curcumin procedure.

Copper, Cobalt, Nickel, Zinc and Molybdenum. Separated by Paper Chromatography and subsequently measured spectrophotometrically after extraction of the bands.

Total Nitrogen. Classical Kjeldahl method.
pH. Determined using the Cambridge pH Meter on soil solution after six hours' contact.
Water-holding Capacity. Saturation of air-dried soil with excess of water, and measurement of the total uptake.
Penetrometry. Measurement of the load needed to penetrate a standard spring-loaded spike to a given depth of soil.

SOIL MONTHLY	pH Water-holding capacity Humus	Available: Phosphate Potash Soda[1] Nitrate[1] Nitrite[1] Ammonia[1]	Available: Iron[1] Copper[1] Cobalt[1] Nickel[1] Zinc[1] Molybdenum[1]	Available: Boron[1] Manganese[1]
SOIL ANNUALLY	Total Nitrogen, Calcium, Magnesium and other mineral substances			
PRODUCE MONTHLY	Total Solids Ash Nitrogen	Calcium Magnesium Potash Phosphate Iron[1]	Copper Cobalt[1] Nickel[1] Zinc[1] Molybdenum[1]	Boron[1] Manganese[1] Vitamin A[1] Vitamin C[1]
PRODUCE ANNUALLY	(Oats, wheat, barley, peas, beans, sugar beet, kale, roots) Total Solids Ash Protein Fat Calcium Magnesium	Potash Soda Phosphate Iron[1] Copper[1] Cobalt[1]	Nickel[1] Zinc[1] Molybdenum[1] Boron[1] Manganese[1]	Vitamin B1[1] Vitamin B2[1] Nicotinic Acid (on cereals)
OTHER PRODUCE ANNUALLY	Compost, FYM, Silage, Hay and Fertilizer Moisture Solids Ash	Nitrogen Phosphate Potash		

[1] Only since 1955.

CROPS

Protein and Mineral Analysis. AOAC methods.
Trace Elements. As for soil (on ashed material).
Vitamins

Vitamin A and Carotene	—Spectrophotometric
Vitamin B_1 (Thiamine)	—Spectrophotometric
Vitamin B_2 (Riboflavin)	—Spectrophotometric
Nicotinic Acid	—Spectrophotometric
Vitamin C	—Titrimetric Indophenol
Vitamin E	—Spectrophotometric.

Arable Crops

Comparisons have been made of the results obtained from the analyses of the crops for protein, fat, ash, minerals (Calcium, Magnesium, Potassium, Sodium, Phosphorus, Sulphur), and the trace minerals Copper, Cobalt, Nickel, Zinc, Molybdenum and Boron, as well as more recently, Vitamins of the B group. Some typical averages are as follows:

VITAMIN B CONTENT		*Average of six years 1955–60*				
		Wheat	*Oats*	*Barley*	*Beans*	*Peas*
Vitamin B_1 (Thiamine)	O	·90	·70	·80	·30	·50
mg./100 g.	M	·80	·70	·70	·40	·50
	S	·80		·70		
Vitamin B_2 (Riboflavin)	O	·12	·15	·15	·14	·13
mg./100 g.	M	·12	·13	·14	·13	·09
	S	·12		·11		
Nicotine Acid mg./100 g.	O	5·20	2·40	9·10	2·80	2·80
	M	5·20	2·40	8·20	2·70	2·80
	S	5·20		8·40		

There is a remarkable consistency about these vitamin findings, with two exceptions, Riboflavin in Mixed Peas being lower than Organic, and Nicotinic Acid in Organic Barley being greatest.

PROTEIN CONTENT

	Average % dry weight for nine years 1952–60				
	Wheat	*Oats*	*Barley*	*Beans*	*Peas*
O	11·5	11·8	11·4	26·9	26·9
M	11·8	11·4	11·6	25·9	26·0
S	12·0		11·2	25·4	

Fodder Analysis (as fed to stock)

	% TOTAL SOLIDS	% Ave.	% ASH CONTENT	% Ave.	% PROTEIN CONTENT	% Ave.
	O	*M*	*O*	*M*	*O*	*M*
Herbage 1956–60	22·60	21·70	2·29	2·22	3·94	3·87
Ley Hay 1952–60	83·60	83·20	6·46	6·34	11·30	10·55
Silage 1952–60	22·00	20·90	3·30	3·72	4·57	4·63

(Organic fodder crops have on average a greater solids content)

VITAMIN A CONTENT—as grazed O 5·12 mg./100 g.
Average of seven months—April–September 1960 M 5·82 mg./100 g.

Cereal Measurements (Average for five years)

	Weight of 1,000 grains in gm.			*Average number of grains per head*		
	Wheat	*Oats*	*Barley*	*Wheat*	*Oats*	*Barley*
O	36·7	35·2	44·3	46·3	34·6	28·6
M	35·1	36·2	42·9	42·6	33·4	27·2
S	34·5	—	42·4	42·0	—	28·1

Concerning our practice of sowing saved seed from the previous harvest, the results tend to indicate that with regard to wheat and oats there is a more noticeable deterioration in the seed grown on the chemical fertilizer sections—but this is not so apparent with barley. We have also observed that, on dressing, the percentage of smalls removed from the Organic cereals is less than that from the Mixed—which in turn is less than from the Stockless cereals.

These tables represent averages from only part of our analytical work. But they give some indication that, although there are many similarities in the analytical findings of the sections, there are also some potentially important differences.

PRODUCE

Comprehensive analysis of the produce from the farms has been made throughout the period under review; towards the end of it more extensive assays for vitamin levels and trace mineral content have been carried out. Reference has already been made to bulk yields of the various crops and to the differences encountered. In addition, the average of some of the analytical figures obtained over the nine-year period are of interest in that this allows of an indication of trends with fluctuations due to field differences and seasonal factors cancelling out. The numbers are too small to allow of satisfactory statistical treatment.

Milk (Average of monthly samples, 1955–60)

	Organic herd	*Mixed herd*
TOTAL SOLIDS	12·50%	12·10%
PROTEIN	3·68%	3·74%

VITAMIN C CONTENT
72 monthly samples were taken;
2 were equal
49 were higher from the Organic herd
21 were higher from the Mixed herd

VITAMIN C CONTENT OF MILK
Monthly variation. Average for the 6 years 1955-1960.

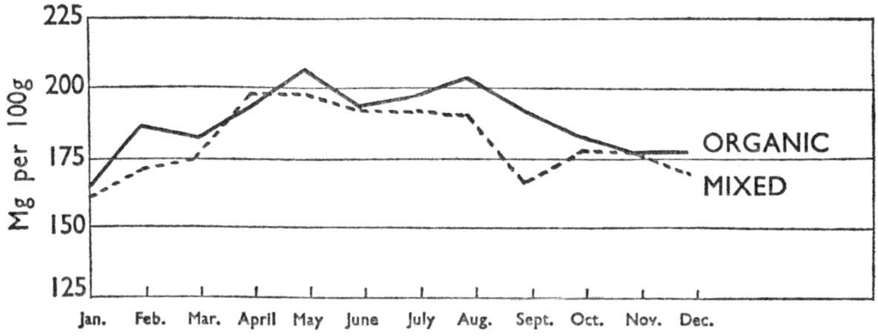

Eggs

Results from 72 samples of eggs from 1955 to 1960 show the following average figures:

	Organic flock	*Mixed flock*
PROTEIN CONTENT	12·5%	12·8%
Average weight of egg	62·5 g.	62·2 g.
Average weight of shell	13·7 g.	12·9 g.

In addition to these analyses regular determinations have been made for ash, mineral content including trace element levels.

Vitamin A Content (Average for the year)

(IU per 100 g.)	*1959: a very dry year* Organic	*1959: a very dry year* Mixed	*1960: a very wet year* Organic	*1960: a very wet year* Mixed
Vitamin A and carotenoids	7,100	6,700	4,300	4,300
Beta Carotene (precursor to Vitamin A)	1,600	1,700	700	700

VITAMIN A CONTENT OF GRAZING. Monthly variation 1960.

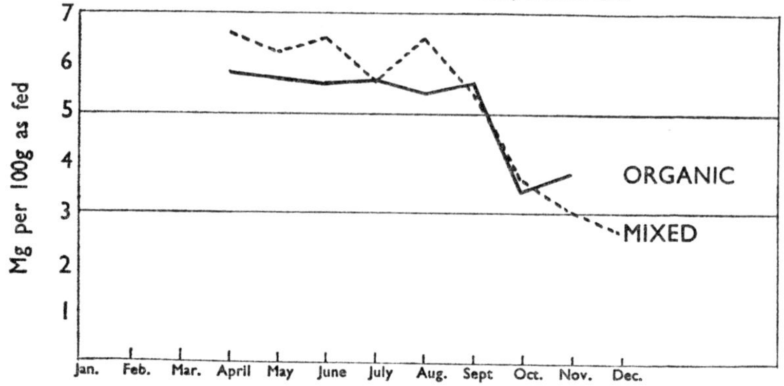

VITAMIN C CONTENT OF GRAZING. Monthly variation 1960.

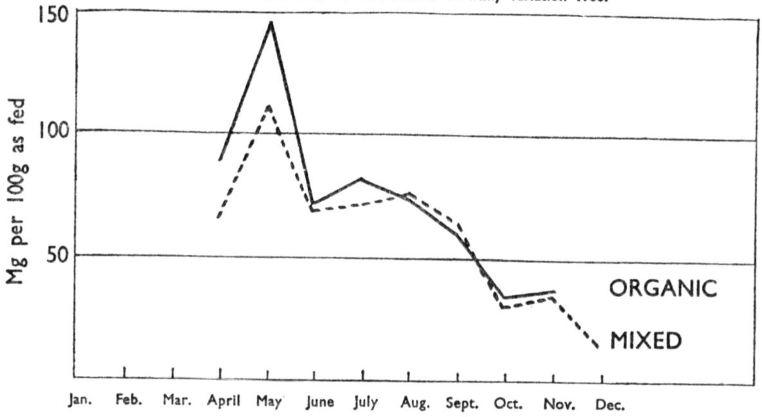

EXPLORING THE UNKNOWN

In concluding our comments on the period we must come back to one major fact emerging: in both health records and quantity of milk, the Organic herd appears to be superior to the Mixed herd, in spite of the substantially greater quantity of food available to the Mixed.

We looked to the analyses of the fodder from the two sections for an explanation but there is almost nothing to account for any marked nutritional difference. It is obvious that we must look further for the explanation. This is what has prompted us to suggest that the time is now ripe for extending the research into possible unknown nutritional factors.

APPENDIX 4

Dr. Milton's Summary of Further Analytical Work Carried Out at Haughley in 1960

The annual report submitted to the Council of the Soil Association on analytical work carried out in connection with the Haughley Experiment during 1960 contains some 150 tables and figures, and represents in all more than 6,000 analyses carried out during the year. The figures derived from these analyses have been studied with a view to correlating differences as a result of the three farming methods used, and conclusions have been arrived at as to the possible significance revealed.

This summary therefore represents a short resumé of the salient features from the Report, and should be read in conjunction with summaries published relating to previous years' work.[1]

In the full report all the details have been given of the routine working of the farm and the various significant factors concerning management including the method of division of the farm into the three sections; the rotational systems applied and of any variation of procedure from the usual pattern which might have been applied during the period under review. However, the bulk of the report is concerned with the analytical results derived from sampling from the soil, crops, produce and animals; details of crop and produce yields (total bulk) and details of chemical fertilizer and/or compost and FYM application.

The design of the analytical work has followed a pattern similar to that of previous years. This is a deliberate attempt to preserve continuity in investigation into the experiment which is itself continuous and dynamic. The year 1960 represented the eighth in the rotational period since full-scale analytical work was inaugurated in 1953, and the results obtained should now be more nearly related to

[1] At present this only applies to members of the Soil Association in whose Journal previous summaries have been published.

the rotational and management pattern imposed upon the fields, than has previously been the case. In other words, the imprint of the farming procedure should now be making itself felt on the three sections.

Our usual method of sampling has been carried out during 1960, viz.:

MONTHLY. Milk, eggs, herbage, soil (each field of the sections following the same diagonal across the field).

YEARLY. All the crops obtained from the farms.

SOIL

A table[1] (of one field in each section) illustrates the scope of the monthly analyses carried out on the soil—and this table is also included so that the reader[2] can compare the results with those obtained in previous years with the same fields, in this way indicating the type of fluctuations which occur from year to year. It is of interest to note that analyses for minor elements have also been made on all fields monthly and these are included for the sake of comparison.

Humus

In our summary of the 1959 experimental work, we pointed out that the tendency for humus levels to rise on the Organic section and to fall on the Stockless section has become less noticeable as the rotations proceeded to completion. The position with regard to the Mixed section lay in between. It now appears that the humus status on the various sections is being maintained. On average the fields on the Stockless section are considerably lower in humus content than on the Mixed section and those on the Organic section are highest of all. Except for an occasional anomalous result, the fields which have consistently showed a low humus level throughout the years continue to do so and likewise those originally high in humus have maintained the order of level, but the downgrading in the Stockless section and the upgrading in the Organic fields appears to have been checked. This could be in accordance with the fact that the rotational system is sufficient to maintain the soil status with regard to fertility. It should be pointed out that the fields on the Organic section only receive compost twice (or possibly once) during the whole of the rotational period and the same applies to the Mixed section with regard to

[1] See p. 323. [2] See footnote, p. 316.

FYM. On the other hand, the two stockbearing sections are re-fertilized by the four-year leys which supply humus to the soil from the grass roots and residues, and by the constant application of animal manures and poultry droppings as the fields are progressively grazed and trodden. This procedure should supply humus in sufficiency, and indeed such would appear to be the case. With the Stockless fields, although these are in the main lower in humus than on the other sections, the use of a ploughed-in green crop on each field twice in the five-year rotational period would seem to have slowed down the rate of humus wastage to a point where the yearly drop in level is barely perceptible.

pH

Quarterly analyses were made for pH on the soils from all fields and we have continued to observe surprising fluctuations throughout the year, e.g. from 7·2 to 8·5 on the same field. We have especially observed for any evidence of trace element deficiency correlated with the highly alkaline conditions, but to date have not noticed such on any of the sections.

Water-holding Capacity and its relation to Humus Content

We continue to note the correlation between this factor and the humus content of the soil.

Thus one Stockless field with humus content of 2·9 per cent had a water-holding capacity of 43 per cent and an Organic field with 5·9 per cent humus had a water-holding capacity of 60 per cent.

There is continuous evidence that those fields of lower water-holding capacity tend to flood after heavy rain whereas those of higher capacity may be seen to have absorbed the water with rapidity under the same circumstances. This factor is also of significance in periods of dry weather whereby the former fields dry out and harden on the surface as opposed to the friable condition which is maintained on the latter. Such factors naturally affect soil structure and it is constantly observed that the low humus fields on the Stockless section show loss of aggregate structure after heavy rain, whereas those of high humus content on the Organic section maintain a good aggregate status under the same conditions. It should be pointed out that a good aggregate size in a sample of soil indicates high fertility in that capillary drainage would be satisfactory and soil aeration will be maintained. It has been pointed out that when a soil reaches a humus level as low as 2 per cent then fertility has virtually disappeared. Although none of the fields on the Stockless section has fallen to this level, a figure of 2·4 per cent has been recorded on some of the fields.

Available Minerals

We continue to record marked fluctuations in the levels of the available minerals throughout the year. On the Mixed and Stockless sections this is largely influenced by the application of chemical fertilizer, although surprisingly enough this effect appears to wear off rather rapidly. We can therefore study the fluctuations on the Organic fields with certainty that any significance will be unrelated to applied fertilizers of any kind (compost excepted). In this case peaks in levels are again seen from May to July, and with a corresponding fall-off towards the winter months. In general, fluctuations are again more noticeable in fields with arable crops and the results from the leys in their second and subsequent years show a more even picture. Thus, in an Organic field of medium humus content with a wheat crop, the level of phosphate in January was 8 mg./100 g. rising to 19 in July and falling to 7 in December. A field under oats gave a phosphate of 12 in January rising to 31 in July and falling again to 7 in December.

We have carried out laboratory tests on such soils using a culture medium containing calcium phosphate which is milky, white and turbid. The plate has been inoculated with a soil suspension and incubated for a period. A clear zone indicates that bacterial production has released phosphate. The spread of the clear zone indicates the degree of phosphate-releasing activity and we have shown that this correlates with the phenomenon of a high-phosphate peak found in practice with the particular soil.

Levels in potash follow much the same pattern as with phosphate, but the fluctuations are less marked.

Nitrate, Nitrogen and Ammonia

We have to report that the levels of nitrate and ammonia on the fields of the Stockless and Mixed sections bear no relationship to the type or quantities of nitrogenous fertilizer added. Variations are prevalent, but these are considerably less than we have found previously, although there is definite evidence of a constant resulting from the product between the two, i.e. as one increases the other decreases. On the Organic fields the levels of both nitrogen fractions are of the same order as found on the other sections where nitrogen fertilizer is applied in inorganic form.

Arising out of this one would expect that the total nitrogen on the Organic fields would be less than on the other sections. In point of fact the reverse is rather the case, and higher results on nitrogen content are found on certain Organic fields—noticeably those of high humus content.

We have calculated that only one-eighth of the nitrogenous fertilizer applied to the Mixed section is removed from the farm in the form of produce (milk and eggs). Since there is no corresponding increase in nitrogen reserve, it would seem that the bulk of applied nitrogen is lost by leaching or by volatilization into the atmosphere.

Minor Elements

We have definite evidence now that variation in availability of certain of the minor elements follows the same pattern as we have previously found for the major elements, nitrogen, potassium and phosphorus. This is particularly so for *manganese*, and the greatest variations are found in fields on the Organic section. Thus (Lane field 'O'), January Manganese 16 ppm, June 60 ppm, and December 27 ppm.

The average levels on all the fields for these three months were:

	January	*June*	*December*
Organic	22 ppm	46 ppm	20 ppm
Mixed	26 ppm	45 ppm	24 ppm
Stockless	30 ppm	32 ppm	21 ppm

Molybdenum and zinc show similar trends, but the figures for copper, cobalt, nickel and boron do not seem to be so influenced. Nevertheless there is no evidence at all that trace element deficiency is present in any of the soils and we have not found any evidence of this in any of the crops so far examined.

CROP RESULTS

(See table on p. 218)

Grain

Highest yields of wheat were from the Mixed fields and lowest from the Organic.

Barley from the Stockless and Mixed sections were similar but that from the Organic section was considerably less. Yield of oats from the Mixed section was very slightly greater than from the Organic field.

Organic peas were higher in yield than Mixed, but the position was reversed with beans and the Stockless beans were highest of all.

Fodder Crops

The yield of Organic maize was heavier than the Mixed, but the yields of Mixed kale, silage and hay were all heavier than the Organic.

There also appears to be more herbage on the Mixed section.

In the cereal crops most of the cereal measurements (see p. 312) were highest from the Organic crops.

Vitamins in Crops

The vitamin B analysis of the cereal crops showed a surprising consistency of levels for aneurine, riboflavin and nicotinic acid from all sections. The same remarks apply to beans and peas.

The vitamin C content of the grazing material showed considerable fluctuations throughout the year. On all leys the highest figures were found in May—gradually falling off to a very low level in December—in some cases being less than one-tenth of the peak figure. The vitamin C levels were generally higher from the Organic leys—but it must be remembered that comparisons are complicated due to the different composition of the ley mixture.

The vitamin A content was less variable than in the case of C, but higher figures were consistently obtained in the summer months—the late autumn and winter samples being less than half those found in April to August.

Protein Content of Herbage

These analyses showed that section differences evened out from one ley to another, but consistently lower figures were given on both sections during the winter months, and these remarks are equally applicable to the mineral content of the grazing material also.

MILK RECORDS

The total quantity of milk was again considerably greater from the Organic section, being 9,203 gallons as against 8,026 from the Mixed herd. From full-time cows and heifers an average of 734 gallons per cow was obtained from the Organic herd as opposed to 579 gallons from the Mixed animals. From part-time animals the figures were 625 and 535 gallons respectively.

Vitamin C content analysis was made monthly. The average figures were:

Organic	2·61 mg./100 g.
Mixed	2·27 mg./100 g.

The fluctuations showed a seasonal pattern, the figures being highest from both sections from May to August, but the figures are considerably higher than we found in the previous year when, due to drought, the grazing was scanty.

Protein content average over the year:

Organic 3·38 per cent *Mixed* 3·52 per cent

Mineral contents. The figures were similar on the two sections.

EGGS

There was very little average difference in chemical composition of the eggs from the two flocks as far as protein, minerals and vitamins were concerned. It is of interest to note that the vitamin A content of the eggs in July was more than double the figure obtained in December.

SUMMARY

There are very many interesting results from the 1960 records, but again the outstanding feature is the considerably higher milk yield from the Organic herd. In 1960 there was adequacy of grazing on both sections, but throughout, there was certainly more keep on the Mixed side in terms of actual grazing bulk, and the rations fed to this herd were more than to the Organic (pp. 220, 305–7). The figures for milk yield per cow are considerably more from the Organic herd when one compares full-time animals only and for this section are higher than in 1959. In the case of the Mixed herd the average per cow is substantially less than in 1959 (a drought year). Thus again we had more milk with less food from the Organic herd, and as the analyses show, the milk is certainly not less in quality as measured by protein, vitamins, minerals, etc. It is, therefore, emphasized that there is some factor in the food given, or in the characteristics of the Organic herd which allows of greater milk production—a finding which has been consistent throughout the experiment to date.

We have to report that the greatest bulk yield of cereals came from the two sections which are given fertilizer. The other arable crops from the Mixed and Organic sections are variable in yield, without any general pattern emerging.

If, however, we consider the cost of the chemical fertilizer added to the Mixed section, then the yield differences did not seem to justify such additions—particularly as the amount removed by the produce (milk and eggs) was not more than a fraction of the NPK applied, and as the excess left no apparent residue in the soil.

The emergence of the fact that availability of certain trace elements seems to follow the same seasonal fluctuating pattern as we have previously seen with phosphorus and potash is very striking—

1960 COMPARISON OF THREE FIELDS

"STABLE"(O) "STACKYARD"(M) "FURTHER BREAK"(S)

		March	*June*	*Sept.*	*Dec.*
pH	O	8·2	8·4	7·8	8·0
	M	7·4	7·5	7·2	7·6
	S	8·1	8·0	7·4	7·6
Humus %	O	6·2	7·8	6·9	6·2
	M	3·7	4·1	3·6	3·3
	S	3·0	3·3	3·1	2·8
Water-holding capacity %	O	62·2	66·7	58·1	62·2
	M	50·8	53·6	49·8	51·1
	S	46·3	49·9	49·6	46·5
Phosphate mg. %	O	27·0	23·0	12·0	19·0
	M	16·0	33·0	24·0	18·0
	S	19·0	29·0	53·0	17·0
Potash mg. %	O	9·2	10·0	12·0	8·2
	M	8·2	8·4	6·8	5·2
	S	10·4	10·4	9·2	7·8
Nitrate Nitrogen ppm	O	1·0	3·3	1·0	2·3
	M	1·5	5·8	1·9	7·6
	S	9·8	7·5	2·0	2·0

particularly as the degree of the fluctuations is greatest in soils with highest accumulated humus content.

As we near the end of the rotational cycle, any differences between the sectional findings must have an increasing significance.

APPENDIX 5

Dr. Milton's Summary of Analytical Work Carried Out in 1961, Together With Certain Rotational Trends

---

The analytical work carried out on the Haughley material during 1961 followed very much the same lines as in previous years—but some considerable additional investigation was made into trace element levels of soils and crops, and more work was carried out on pasture content and behaviour.

During 1961 the monthly soil sample from each field has been analysed for:

pH
Water-Holding Capacity
Humus
Available Phosphate
Available Potassium
Available Sodium
Extractable Nitrate N
Extractable Ammonia N
Available Manganese
Available Magnesium
Available Molybdenum
Available Nickel
Available Copper
Available Cobalt
Available Zinc
Available Boron
Available Iron

Crops have been analysed for content of major and minor elements, and vitamins, as well as for moisture, protein, cellulose, etc.

Produce (milk and eggs) has also been analysed in the same manner.

In addition, the crops have been tested for radioactivity levels.

Nitrogen levels have been given special consideration, and on all the crops we have analysed not only for total nitrogen, but also for non-protein nitrogen and amino acid nitrogen.

The protein nitrogen has been further broken down and we have obtained figures for soluble protein glutenin and gliadin on all the crops.

Preliminary work has also been carried out on the differential sugar content of the crops, and we have obtained results for total sugars in all cases.

We have also analysed for oil content of the crops, but have not yet carried out differential fat analysis.

Vitamin determinations (B_1, B_2, Nicotinic acid and total tocopherol) have been made on all the crops. During the 'growing' months, levels of vitamins A and C have been determined monthly on the herbage.

Considerable attention has been paid to the monthly analysis of herbage. Besides the above-mentioned vitamins, tests have been made for nitrogen, ash and moisture.

The monthly milk analysis has been continued, and this has included nitrogen, potash and phosphate, and determinations of vitamin A, B, Carotene and vitamin C levels.

The monthly egg analysis has followed the same pattern as for milk.

Some attempt to assess the results obtained has been made. It will be of interest to consider both the results of the 1961 analyses in isolation, and the rotational trend.

Water-holding Capacity

The apparent correlation between the humus content and ability of the soil to retain moisture seems to be a feature of the work. Other factors clearly play a part, but certainly the results show broadly that land with high humus content is better able to withstand drought and less likely to become waterlogged after heavy rain than soil less richly endowed.

The rotational picture, wherein the figures from 1952 to 1961 are compared, shows very interesting trends, i.e. on all the sections the WHC has increased during this period to a definite extent. The results on deeper investigation indicate (as expected) that a second factor in WHC is concerned with the amount of clay material in the soil—the figures on lighter lands being more closely correlated to the humus content than is the case with the heavy clay soil.

WATER-HOLDING CAPACITY RELATED TO HUMUS CONTENT OF SOIL

Humus	*Water holding capacity*		
content	*Organic*	*Mixed*	*Stockless*
3·0–3·5%	55	52	46
3·5–4·0%	53	53	47

pH

The 1961 figures show the now expected variation of a seasonal nature—in some of the Stockless fields this can vary from perhaps 6·4 to 8·5, and is probably associated with fertilizer placement. The fluctuations on the stockbearing sections are less marked.

The rotational trend suggests that all the fields on the farms are becoming less alkaline.

SOIL ACIDITY (pH)
(Average pH of all fields analysed)

Year	*Organic*	*Mixed*	*Stockless*
1952	8·42	8·33	8·32
1953	8·35	8·22	8·16
1954	8·07	7·95	7·87
1955	7·96	7·93	7·87
1956	7·77	7·92	8·00
1957	7·90	7·81	7·79
1958	7·90	7·78	7·73
1959	8·03	7·91	7·81
1960	7·96	7·62	7·83

Humus

The 1961 figures indicate the now expected fluctuations in humus content.

The rotational picture now suggests that there has been a definite increase on all the sections, and that this trend is continuing as can be seen from the averages shown in the tables on p. 328, and this has been more marked since 1958.

HUMUS CONTENT OF SOIL

	1953–5	*1954–6*	*1955–7*	*1956–8*	*1957–9*	*1958–60*
Organic	3·16	3·29	3·38	3·41	3·37	3·62
Mixed	2·90	3·06	3·22	3·27	3·18	3·30
Stockless	2·67	2·75	2·81	2·84	2·77	2·93

RISE IN HUMUS CONTENT OF 1958–60 OVER 1953–5

Organic	14%
Mixed	14%
Stockless	10%

The humus levels appear from the rotational figures to be highest in the months April to June, and no apparent reason can be attributed to this fact. If, however, we take the humus levels under the different crops (spring three-month averages) a very interesting graph results as we follow the rotations on the stock-bearing sections.

Under ley the humus level on the Mixed section rises rapidly—with an equally sharp fall as the field is under cereals. There is another rise under pulses, and a second fall with wheat and silage.

With the Organic section, the rises and falls are identical but less marked, although it becomes obvious at what stage compost is applied, as it is (but to a lesser extent) when FYM is applied to the Mixed section. (See Graph 7.)

GRAPH 6

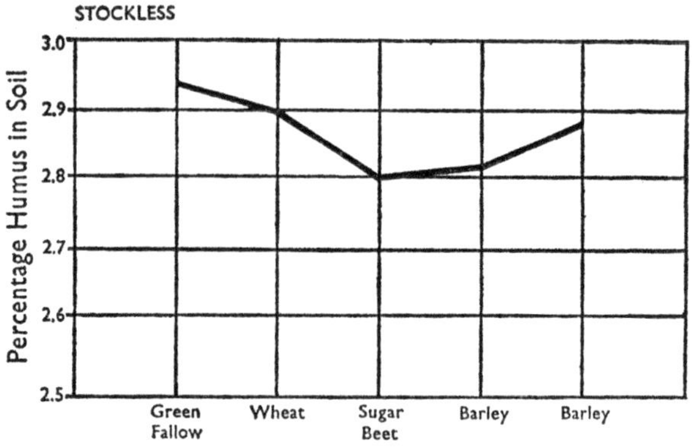

With the Stockless rotation the fluctuations in humus levels are much less marked—there being a very small variation from crop to crop; this despite the large amount of ploughed in organic residues, clover, trefoil, beet tops, barley stubble and straw. (See Graph 6.)

GRAPH 7

HUMUS CONTENT OF SOIL %. Average three monthly samples April, May, June: Eight years 1953-1960. For most of this period the rotation was in the order shown.

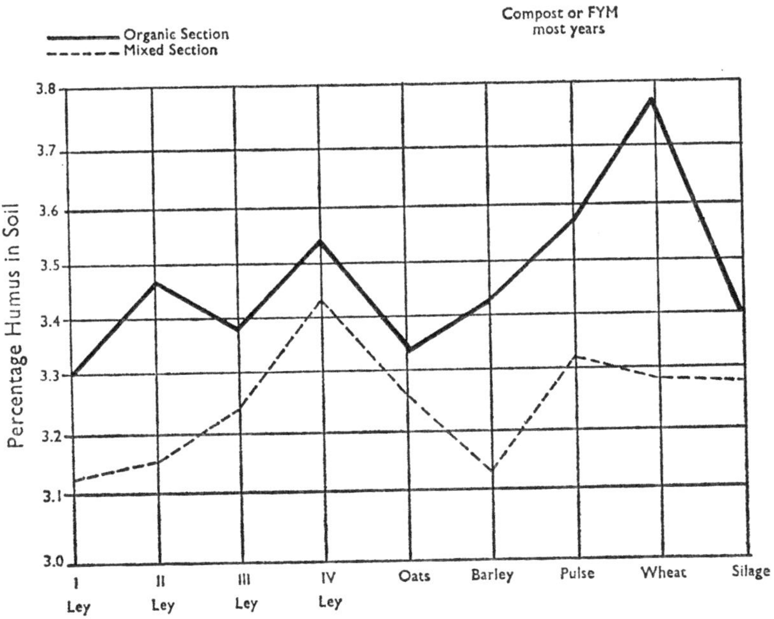

For Graphs of

Available Nutrients

See pp. 227–8

THE EFFECT OF CHEMICAL FERTILIZER APPLICATION ON QUANTITY OF NUTRIENTS REMOVED IN CROP

Average for the Years 1953-60

On the Organic and Mixed sections these nutrients are returned to the section through the stock, removals taking place only through sales of animals and animal products. On the Stockless section they are sold off.

AVERAGE NUTRIENTS REMOVED AND APPLIED ON THREE SECTIONS

(cwt./acre)

WHEAT		ORGANIC	MIXED	STOCKLESS	
Nitrogen	Removed	0·40	0·47	0·42	
	Applied	—	0·35	0·50	
Phosphate	Removed	0·22	0·25	0·22[1]	
	Applied	—	0·74	0·64[1]	
Potash	Removed	0·12	0·13	0·11	
	Applied	—	0·55	0·57	
OATS					
Nitrogen	Removed	0·42	0·37	Not grown	
	Applied	—	0·12		
Phosphate	Removed	0·19	0·16	Not grown	
	Applied	—	0·23		
Potash	Removed	0·11	0·10[1]	Not grown	
	Applied	—	0·11[1]		
BARLEY				*After Sugar Beet*	*After Barley*
Nitrogen	Removed	0·41	0·38	0·44	0·36
	Applied	—	0·24	0·40	0·39
Phosphate	Removed	0·26	0·20	0·25	0·20
	Applied	—	0·31	0·56	0·45
Potash	Removed	0·15	0·15	0·17	0·14
	Applied	—	0·28	0·42	0·40

[1] Omitting 1955.

AVERAGE TRACE ELEMENTS ON SECTIONS

(all as ppm)

	Stockless	*Mixed*	*Organic*
Magnesium	24·000	26·000	24·000
Manganese	48·000	50·000	51·000
Molybdenum	0·780	0·690	0·880
Copper	2·490	2·530	2·690
Nickel	0·790	0·740	0·770
Cobalt	0·770	0·790	0·900
Iron	145·000	161·000	174·000
Zinc	1·180	1·240	1·260
Boron	0·065	0·066	0·072

Trace Elements Content of Crops, Milk and Eggs

The levels in all crops of copper, cobalt, nickel, zinc, molybdenum, manganese, iron and boron have been assessed. The differences between the crops from the three sections are quite small, and on the basis of one year's results, we do not think that such differences are significant.

The levels in milk suggest that material from the organic herd is higher in copper, cobalt, nickel and zinc than that from the mixed herd, when results are averaged over the year.

Eggs: no significant differences have been observed.

Protein Content of Crops

Although the bulk yield of the organic cereals is usually less than that from the other sections, the trend seems to be for protein content to be highest from the Organic section. In 1961 this was true for the barley and wheat, but not for oats. The protein from the cereals of the stockbearing sections is usually higher than that from the Stockless section.

Non-Protein and Amino Acid Nitrogen

No apparent sectional differences have been observed.

TRACE ELEMENTS IN CROPS

Crop		*ppm on samples dried at 100° C.* B.	Cu.	Co.	Ni.	Zn.	Mo.	*Percentage mg. dried* Mn.	Fe.	Ca.	Mg.
BARLEY											
Pear Tree	S	0·1	1·0	1·0	0·2	4·0	0·1	19	50	60	150
Further Break	S	0·2	1·0	1·0	0·4	4·0	0·1	20	50	60	160
	M	0·1	2·0	1·0	0·1	4·0	0·1	25	60	40	150
	O	0·1	2·0	1·0	0·2	4·0	0·2	28	60	40	150
WHEAT											
Little	S	0·1	1·0	1·0	0·8	4·0	0·1	18	50	80	120
Road	S	0·1	2·0	1·0	0·2	4·0	0·2	20	60	50	110
	M	0·2	3·0	1·0	0·6	4·0	0·1	19	50	60	120
	O	0·1	2·0	1·0	0·4	3·5	0·1	28	40	50	160
Heat-treated 1	O	0·1	2·0	1·0	0·4	4·0	0·1	22	50	40	150
Heat-treated 2	O	0·1	3·0	1·0	0·8	4·0	0·1	18	30	50	160
OATS											
Spring	M	0·1	1·0	1·0	0·4	4·0	0·1	21	50	50	190
Winter	M	0·1	2·0	1·0	0·2	4·0	0·1	28	50	60	210
	O	0·2	2·0	1·0	0·2	4·0	0·1	24	50	40	190
BEANS	S	0·1	4·0	1·0	0·6	4·0	0·1	30	60	90	240
	M	0·1	2·0	2·0	0·2	4·0	0·1	23	50	40	160
	O	0·1	4·0	2·0	0·2	4·0	0·1	22	60	80	160
PEAS	M	0·1	1·0	1·0	0·1	3·0	0·1	25	50	50	140
	O	0·1	4·0	1·0	0·2	4·0	0·2	25	60	80	180

Soluble Protein, Glutenin and Gliadin

Soluble protein and gliadin follow the same pattern as for total protein, i.e., the crops from the stockbearing sections are higher than from the Stockless farm. With glutenin the reverse situation is apparent.

Differential Sugar Analyses

Preliminary investigation of the 1961 crop indicate possible differences in sugar content from the different sections. With wheat and barley it would appear that the maltose content is significantly higher from the Organic section, but no differences were observable in the oats, peas and beans from the various sections.

Oil Content of Crops

Sectional differences were not significant.

Vitamin Levels of Crops

Analyses on all crops have been made for thiamine, riboflavin, nicotinic acid (all B group) and total tocopherols (vitamin E).

Sectional differences were not significant.

VITAMIN ANALYSES OF PASTURES

VITAMIN A mg./100 g. as received

	March	*April*	*May*	*August*	*September*
Mixed	5·37	5·72	5·95	5·20	4·30
Organic	4·90	4·62	4·05	4·56	4·86

VITAMIN C mg./100 g. as received

	March	*April*	*May*	*August*	*September*
Mixed	81	81	65	81	99
Organic	71	107	75	121	149

PROTEIN CONTENT OF PASTURES

(*percentage of dry weight*)

	March	*April*	*May*	*August*	*September*
Mixed	13·5	13·8	11·6	13·3	15·2
Organic	13·2	12·6	13·6	15·4	16·0

TOTAL MINERAL CONTENT OF PASTURES

(percentage of sample dried at 100° C.)

	March	*April*	*May*	*August*	*September*
Mixed	12·2	12·8	13·1	11·7	11·0
Organic	12·4	12·1	12·4	13·3	12·8

TOTAL MOISTURE CONTENT OF PASTURES

(percentage of samples as received)

	March	*April*	*May*	*August*	*September*
Mixed	82·2	81·1	76·4	77·5	77·0
Organic	80·8	80·8	76·4	74·6	80·1

VITAMIN C CONTENT OF MILK

(mg. percentage)

	Jan.	*Feb.*	*Mar.*	*Apr.*	*May*	*June*	*July*	*Aug.*	*Sep.*	*Oct.*	*Nov.*	*Dec.*
Mixed	2·00	2·15	2·40	2·30	3·40	3·00	2·80	2·50	2·59		2·90	2·90
Organic	2·00	2·30	2·50	2·80	3·10	2·90	3·00	2·00	2·40		3·00	3·08

VITAMIN A CONTENT OF EGGS

IU/100 g.

	Jan.	*Feb.*	*Mar.*	*Apr.*	*May*	*June*	*Aug.*	*Oct.*	*Nov.*	*Dec.*
VITAMIN										
Mixed	4,600	4,200	3,800	5,200	3,500	5,800	6,200	5,000	3,900	3,500
Organic	3,900	4,400	6,100	5,400	5,100	4,000	3,600	4,400	4,500	3,900
BETA CAROTENE										
Mixed	650	760	690	900	880	710	690	710	740	680
Organic	650	740	740	900	890	900	760	770	820	730

General Statement

The above tables and summaries of the 1961 analyses are in the main made without comment. Interpretation of the results show that there are a number of leads wherein our search for 'quality' in foodstuffs might be pursued.

The summaries of some average results for soils taken over a rotational cycle make interesting reading. We are now at the stage where our soil analyses are being statistically treated, and the exact rotational changes in soil pattern should be revealed accordingly.

MISCELLANEOUS ANALYSES

		Percentage on sample as received					Percentage on sample dried at 100° C.				
Crop	H_2O	*Ash*	*N*	P_2O_5	*K*	*Na*	*Ash*	*N*	P_2O_5	*K*	*Na*
SILAGE:											
OATS AND PEAS											
M	69·4	1·20	0·56	0·15	0·23	0·10	3·90	1·82	0·49	0·76	0·32
O	84·4	1·02	0·27	0·09	0·15	0·06	6·70	1·75	0·59	1·00	0·38
LUCERNE											
M	82·5	1·10	0·32	0·13	0·17	0·06	6·30	1·80	0·75	0·98	0·32
O	76·0	1·73	0·48	0·14	0·24	0·10	7·20	1·99	0·59	0·98	0·41
HAY											
M	17·1	6·10	2·78	0·62	0·80	0·09	7·40	3·36	0·75	0·97	0·11
O	14·3	7·00	3·05	0·60	0·78	0·16	8·20	3·59	0·71	0·92	0·19
MUCK											
M	86·5	3·20	0·28	0·15	0·15	0·08	23·6	2·10	1·12	1·08	0·60
COMPOST											
O											
Bushey E.	67·0	8·40	0·44	0·45	0·36	0·22	24·8	1·30	1·33	1·06	0·65
Nappers	76·0	9·60	0·31	0·32	0·29	0·15	39·7	1·30	1·32	1·22	0·61
Stable	69·4	5·64	0·40	0·33	0·39	0·20	18·4	1·30	1·08	1·26	0·64

APPENDIX 6

Dr. Milton's Summary of Analytical Work Carried Out in 1962 and 1963

During 1962 and 1963 more than 150 tables of figures have been compiled. These figures are the basis for the study of ascertaining differences arising from the three farming methods we use and the conclusion which may be derived therefrom.

The following is a brief summary of the 1962 and 1963 reports. It should be read in conjunction with summaries printed in the two previously published reports.

During 1962 and 1963 soil analysis, of which the major portion of the analytical work consisted in past years, has been considerably curtailed, and the comprehensive analysis, previously carried out monthly, on each of the fields has been made in May and September only.

The figures obtained on these occasions have been averaged throughout the sections. These figures show that the humus content on the Organic section is almost 1 per cent higher than on the Mixed section, and almost 2 per cent higher than on the Stockless section. On these averages also the Organic section has the highest total nitrogen despite the lack of application of nitrogenous fertilizers. The nitrate nitrogen, ammonium nitrogen and available phosphate are all higher on the Organic section, and the potassium is higher than on the Mixed although not so much as on the Stockless.

Although one must be very cautious about the interpretation of average figures, the picture is sufficiently clear to indicate that on the Organic section, which is not treated chemically, there is no nutritional deficiency. In point of fact, as is shown by the total nitrogen, the general level is usually higher in nutrient substances.

It is proposed in future to follow this pattern of twice yearly sampling of the soil, and it may be of interest to take the samples in late June and early January, which conforms to the peak of growth and the trough of quiescence in the natural cycle of things in the soil.

In place of the work previously done on soil analysis, the laboratory's activities have been directed into other channels. In particular, very much more work has been done on pasture analysis. All pastures have been analysed from May to September inclusive. This works out to well over 1,000 analyses on pasture alone. Some of the techniques concerned in this work are very complicated, and give an indication of the considerable amount of time which has been given to this branch of the investigation.

It can reasonably be stated that we have no evidence that pastures on the Mixed section contain a higher proportion of free nitrate than those on the Organic section; nor are we able to obtain much indication of the differences of vitamin content or trace element content by sectional comparison.

We have extended our work on the analysis of milk. This now includes monthly analyses of trace elements, and we have also made regular differential analyses of the fatty acids in the milk. Similarly, extension of the work on eggs has followed the same line.

The work on the analysis of crops has also been considerably widened. The 1962 work includes trace elements in straws, and more comprehensive cereal measurements. In addition to the determination of vitamin B group in crops, we have also included total tocopherols, i.e. vitamin E, in these substances. It is of interest to note that in 1962 the vitamin E content of barley was considerably higher than that of the other cereal crops (see table on p. 344). In previous years nicotinic acid was also greatest in barley but our 1962 analysis showed a level of this vitamin not dissimilar from that of the other cereals. besides the normal comprehensive analysis we have also made a determination of vitamin C content of maize. The total fat content of all crops has also been determined.

The vitamin analysis on the 1963 crops shows some changes, e.g. in the case of wheat, vitamin B_2 appears higher, and vitamin E is lower. With barley there is a consistent increase in B_2 level, and a drop in vitamin E level. With oats again there is a higher level in vitamin B_2. The beans show increase in vitamin E level, as is also the case with the peas. These changes may have been due, to some extent, to weather conditions at the time of harvest, as well as to the expected seasonal variation.

Perhaps some of the most interesting extensions to our work have been the investigations into the composition of the sugars in the crops, and we hope to be able to extend this work in future to a quantitative basis.

Whereas in the past we have been concerned mainly with the crude nitrogen content of crops, we have attempted to differentiate the protein fractions, and in particular to determine the non-protein fraction of the cereals.

We have extended our work on muck and compost analysis to include trace element contents so as to complete the story on all the substances concerned in the cycle.

As 1962 was the first year of a more comprehensive attack in order to obtain leads concerning quality in foodstuffs, it is a little premature to give opinions as to results, particularly in view of the fact that there were a number of crop failures during that year.

Our future work will be concerned with following up more refined analyses of crops and pastures. Considerable headway has already been made with methods for determining the amino acid differentiation in the products.

In 1962 we made a start on the gas chromatographic separation of the fatty acids in the crops. Some forty-five fatty acids have been followed up on each of the crops analysed on each of the sections, and it is proposed eventually to make a statistical analysis and interpretation of the results obtained. There are some very striking differences in the percentage content of the various fatty acids in the products from the various sections. We have had some confirmation of this from the work now being carried out on the 1963 crops, and on the material from the fistulated heifers (see below), but it is still too early to comment.

It is of great interest to be able to compare results for the two years over such a wide range.

It certainly cannot be said that any deficiencies are apparent in the produce from the Organic section as a result of not applying chemical nutrients in artificial form.

It would not be proper to finish these observations without commenting on the fact that once again the milk yield from the Organic cows was considerably higher than that from the Mixed herd, and reference to the crop yields shows that there was considerably less feed produced on the Organic section.

It becomes increasingly obvious, therefore, that we must investigate the problem very fully from all aspects in order to clarify the reasons for this consistent difference and to show whether it is in any way related to the composition of the foodstuffs.

OBSERVATIONS CONCERNING MONTHLY SOIL ANALYSES—1962 AND 1963 AVERAGES

	STOCKLESS		MIXED		ORGANIC	
	1962	*1963*	*1962*	*1963*	*1962*	*1963*
pH	7·7	8·0	7·6	8·1	7·7	8·2
(Range)	7·5–8·0	7·4–8·4	7·2–8·0	7·3–8·6	7·4–8·1	7·6–8·5
Humus (g./100 g.)	2·90	2·91	3·92	3·39	4·95	4·47
Total Nitrogen (g./100 g.)	0·23	0·23	0·24	0·23	0·27	0·29
C/N Ratio	12·7/1	12·7/1	16·4/1	14·7/1	18·3/1	15·4/1

AVAILABLE TRACE ELEMENTS IN SOIL, 1962 AND 1963 AVERAGES

	STOCKLESS		MIXED		ORGANIC	
	1962	*1963*	*1962*	*1963*	*1962*	*1963*
Magnesium (mg./100 g.)	20·00	37·00	26·00	36·00	32·00	39·00
Manganese (ppm)	36·00	52·00	39·00	51·00	37·00	50·00
Molybdenum (ppm)	0·60	0·90	0·60	1·50	0·60	1·20
Nickel (ppm)	0·80	0·80	0·80	0·80	0·80	0·70
Copper (ppm)	4·50	3·60	4·00	4·00	4·50	4·60
Cobalt (ppm)	0·75	0·50	0·75	0·60	0·75	0·70
Zinc (ppm)	3·50	2·80	3·00	0·39	3·50	3·80
Iron (ppm)	136·00	226·00	133·00	245·00	174·00	240·00
Boron (ppm)	0·07	—	0·07	—	0·08	—

ANALYSES OF PASTURES 1962

	May	June	July	August	September
VITAMIN A mg./100 g. as received					
Mixed	2·65	3·15	3·42	3·57	3·30
Organic	2·35	2·51	2·82	3·12	2·50
VITAMIN C mg./100 g. as received					
Mixed	102	125	118	124	118
Organic	94	105	114	121	111
PROTEIN CONTENT % of dry weight					
Mixed	20·8	21·4	21·9	21·3	21·0
Organic	22·2	21·5	20·7	22·5	21·3
TOTAL MINERAL CONTENT % of sample dried at 100° C.					
Mixed	10·7	10·6	9·5	11·0	9·7
Organic	10·3	10·6	10·4	11·4	12·2
TOTAL MOISTURE CONTENT % of sample as received					
Mixed	83·2	79·9	80·0	81·8	84·2
Organic	81·2	77·5	76·3	78·9	84·3

ANALYSES OF PASTURES 1963

	June	August	October
VITAMIN A mg./100g. as received			
Mixed	6·5	7·3	7·2
Organic	6·2	6·4	6·7
VITAMIN C mg./100 g. as received			
Mixed	98	126	116
Organic	107	101	113
PROTEIN CONTENT % of dry weight			
Mixed	16·8	17·4	17·0
Organic	16·1	16·9	15·7
TOTAL MINERAL CONTENT % of sample dried at 100° C.			
Mixed	9·4	10·2	12·5
Organic	8·7	10·5	12·2
TOTAL MOISTURE CONTENT % of sample as received			
Mixed	81·6	80·5	85·3
Organic	77·0	73·0	80·3

Dry Matter % of the first-year leys during May 1963

Section	*Organic*	*Mixed*	
Name of field	Wood (8½ acres)	Sawpit (8¼ acres)[1]	
		Fertilized[2]	Unfertilized strip
Date of sampling:			
2.5.63	18·2	11·8	13·0
6.5.63	26·0	13·0	13·0
14.5.63	21·8	13·3	12·6
22.5.63	21·5	14·4	13·5

[1] Whole field received 48 cwt. slag on 10.1.63.
[2] All except a centre strip of 1 acre received 20 cwt. F52 on 10.4.63.

MILK ANALYSES 1962 AND 1963

	MIXED 1962	MIXED 1963	ORGANIC 1962	ORGANIC 1963
The variations throughout the year found on the monthly analyses:				
(as %)				
Moisture	85·9–88·0	86·3–88·7	86·0–88·0	86·3–88·4
Solids	14·1–12·0	11·3–13·7	14·0–12·0	11·6–13·7
Ash	0·55–0·80	0·62–0·73	0·58–0·79	0·57–0·71
Nitrogen	0·51–0·64	0·64–0·68	0·52–0·66	0·66–0·69
Phosphate	0·25–0·30	0·26–0·29	0·27–0·30	0·27–0·29
Potassium	0·12–0·19	0·12–0·14	0·12–0·19	0·12–0·14
Averages for the year:				
Solids	13·30	12·70	13·40	12·60
Ash	0·67	0·69	0·68	0·66
Nitrogen	0·60	0·65	0·60	0·67
TRACE ELEMENTS IN MILK				
(ppm)				
Monthly variations:				
Copper	0·20–2·80	0·30–0·80	1·00–1·50	0·30–0·80
Cobalt	0·50	0·30	0·50	0·30
Nickel	0·50–1·80	0·30–0·60	0·50–1·40	0·30–0·50
Zinc	0·50–0·75	0·20–0·80	0·50–0·80	0·30–0·80
Molybdenum	0·10–0·40	0·20–0·30	0·10–0·20	0·20–0·30
Manganese	1·0	0·20–0·30	1·0	0·20–0·40
Iron	8·00–17·00	10·00–26·00	10·00–17·00	13·00–26·00
FAT IN MILK				
(as %)				
Monthly variations:				
Morning	[1]	4·20–5·00	[1]	4·70–5·50
Afternoon	[1]	5·10–6·30	[1]	5·30–6·50
Averages for the year:				
Morning	[1]	4·80	[1]	5·00
Afternoon	[1]	5·40	[1]	5·60
VITAMIN C IN MILK				
(mg./100 ml.)				
Monthly variations	2·23–2·70	2·10–2·60	2·29–3·08	2·20–2·80
Average	2·51	2·40	2·55	2·50

[1] Figures not available for 1962.

These figures show that the organic milk, although greater in quantity, is in no way inferior in quality.

EGG ANALYSES 1962 AND 1963

The regular monthly figures for weight, ash and major nutrients are given. Variations are:

	MIXED *1962*	*1963*	ORGANIC *1962*	*1963*
	grams	*grams*	*grams*	*grams*
Weight	61·000–67·400	62·200–65·500	60·500–67·200	60·400–65·000
	%	%	%	%
Shell	10·700–12·700	9·800–12·300	10·500–12·700	9·500–12·600
Ash	0·640– 0·960	0·720– 0·930	0·790– 0·910	0·640– 0·930
Nitrogen	2·080– 2·380	2·160– 2·240	2·040– 2·310	2·180– 2·280
Phosphate	0·310– 0·500	0·390– 0·510	0·310– 0·470	0·380– 0·490
Potash	0·140– 0·190	0·130– 0·190	0·140– 0·190	0·130– 0·200
Sodium	0·120– 0·150	0·110– 0·150	0·110– 0·140	0·110– 0·150
Calcium	0·052– 0·062	0·052– 0·060	0·053– 0·062	0·055– 0·062
Magnesium	0·007– 0·009	0·006– 0·009	0·007– 0·009	0·006– 0·009

TRACE ELEMENTS

Monthly variations were:

	ppm	*ppm*	*ppm*	*ppm*
Copper	1·0–5·0	0·6–1·6	1·0–5·0	0·6–1·6
Cobalt	0·5	less than 0·3	0·5	less than 0·3
Nickel	0·5–2·0	0·6–1·3	0·5–3·0	0·6–0·9
Zinc	1·0–3·0	0·9–1·6	1·0–3·0	0·9–1·6
Molybdenum	0·5–1·5	0·3–0·6	0·5–1·0	0·3–0·6
Manganese	1·0	0·8–1·4	1·0	0·8–1·3
Iron	75–175	52–85	80–165	65–83

The figures for the above average out as follows:

Average weight	63·60 g.	63·80 g.	64·30 g.	64·00 g.
Shell	11·70 g.	10·80 g.	11·70 g.	10·80 g.
Ash	0·87%	0·82%	0·86%	0·84%
Nitrogen	2·21%	2·20%	2·22%	2·23%

CEREAL MEASUREMENTS 1962

CROP AND SECTION		GRAINS PER HEAD			*Weight of 100 grains*
Wheat		*Max.*	*Min.*	*Ave.*	*in grams*
Chestnut W.	S	65	18	43	3·480
Gypsy	M	62	29	41	3·681
Mid. Wassex (heated seed)	O	69	44	54	4·235
Mid. Wassex (dressed seed)	O	65	24	44	4·061

CROP AND SECTION		GRAINS PER HEAD			*Weight of 100 grains*
Barley		*Max.*	*Min.*	*Ave.*	*in grams*
Pear Tree	S	30	17	25	4·247
Chestnut E.	S	29	21	28	4·047
Stackyard	M	30	21	26	4·060
Nappers	O	29	21	29	4·230
Stable	O	34	24	26	4·481
Oats					
Pond (spring)	M	40	20	33	3·521
Pond (winter)	M	41	20	34	3·462
Bushey W.	O	43	24	39	3·613

CEREAL MEASUREMENTS 1963

CROP AND SECTION		GRAINS PER HEAD			*Weight of 100 grains*
Wheat		*Max.*	*Min.*	*Ave.*	*in grams*
Further Break	S	49	14	33·0	3·721
Big Middle	M	55	13	30·8	3·948
L. Wassex N.	O	41	19	34·8	3·604
Garden	O	38	10	28·1	3·507
Barley					
Chestnut E.	S	27	21	22·1	4·621
Little	S	26	20	23·4	4·647
Road	S	27	11	20·3	4·948
New Bells Long	M	28	19	23·7	4·344
First Break	O	33	17	27·0	4·044
Oats					
7 Acre (winter)	M	39	19	30·9	3·533
7 Acre (spring)	M	32	13	23·2	3·818
L. Wassex S. (winter)	O	69	14	27·5	3·681
L. Wassex S. (spring)	O	68	14	25·5	3·949
Garden (spring)	O	66	14	37·5	3·470
Dredge Corn					
Barley—L. Wassex N.	O	32	21	24·7	4·550
Wheat—L. Wassex N.	O	63	8	35·4	3·590

VITAMIN ANALYSIS OF CROPS 1962

CROP AND SECTION		*mg percentage on samples as received* B_1	B_2	*Nicotinic acid*	*Vitamin E*
Wheat					
Chestnut W.	S	0·53	0·1	3·44	1·99
Cottage	M	0·53	0·1	3·64	2·62
Gypsy	M	0·67	0·2	3·80	2·62
Mid. Wassex (heated seed)	O	0·80	0·1	3·80	3·28
Mid. Wassex (dressed seed)	O	0·67	0·3	3·72	2·62
Barley					
Pear Tree	S	0·67	0·1	3·48	3·94
Chestnut E.	S	0·67	0·3	3·80	3·94
Stackyard	M	0·53	0·2	3·64	4·58
Nappers	O	0·53	0·2	3·48	5·22
Stable	O	0·53	0·2	3·64	5·22
Oats					
Pond (spring)	M	0·53	0·1	3·76	1·99
Pond (winter)	M	0·80	0·2	3·00	2·62
Bushey W. (spring)	O	0·53	0·2	3·16	1·99
Bushey W. (winter)	O	0·67	0·3	3·79	2·62
Beans					
Pear Tree	S	0·67	0·2	4·36	3·94
7 Acre	M	0·67	0·3	4·44	4·58
L. Wassex S.	O	0·53	0·2	4·12	4·58

VITAMIN ANALYSIS OF CROPS 1963

CROP AND SECTION		*mg. percentage on samples as received* B_1	B_2	*Nicotinic acid*	*Vitamin E*
Wheat					
Further Break	S	0·53	0·20	3·60	1·30
Big Middle	M	0·53	0·20	3·74	2·60
L. Wassex N.	O	0·66	0·25	3·60	1·30
Garden	O	0·60	0·25	3·82	2·60
Barley					
Chestnut E.	S	0·53	0·30	3·46	1·30
Little	S	0·53	0·30	3·66	2·60
Road	S	0·60	0·30	3·40	1·30
New Bells Long	M	0·47	0·30	3·60	2·60
First Break	O	0·53	0·40	3·74	1·30

CROP AND SECTION		*mg. percentage on samples as received* B_1	B_2	*Nicotinic acid*	*Vitamin E*
Oats					
7 Acre (winter)	M	0·66	0·30	4·06	2·60
7 Acre (spring)	M	0·80	0·20	3·66	3·72
L. Wassex S. (spring)	O	0·73	0·20	3·82	2·60
L. Wassex (winter)	O	0·80	0·25	3·90	3·72
Garden (spring)	O	0·66	0·30	4·06	2·60
Beans					
Chestnut E.	S	0·66	0·15	3·06	5·20
Stackyard	M	0·66	0·30	4·25	6·50
Nappers	O	0·60	0·30	4·10	6·50

PROTEIN ANALYSIS OF CROPS 1962

Percentage as received

CROP AND SECTION		*Total protein*
Wheat		
Chestnut W.	S	11·60
Cottage	M	10·60
Gypsy	M	13·90
Mid. Wassex (heated seed)	O	8·30
Mid. Wassex (dressed seed)	O	8·70
Barley		
Pear Tree	S	9·20
Chestnut E.	S	9·40
Stackyard	M	9·20
Nappers	O	9·00
Stable	O	7·70
Oats		
Pond (spring)	M	8·00
Pond (winter)	M	11·90
Bushey W. (spring)	O	13·50
Bushey W. (winter)	O	11·80
Beans		
Pear Tree	S	19·20
7 Acre	M	18·80
L. Wassex S.	O	18·00
Peas		
7 Acre	M	15·30
L. Wassex S.	O	Failed

Percentage as received

CROP AND SECTION			*Total protein*
Silage—Lucerne			
Walnut Tree Long W.		M	2·54
Lane		O	2·46
Silage—Oats and Peas			
Sawpit		M	3·00
Wood		O	2·37
Hay—Lucerne			
Walnut (T.L.) W.	2nd cut	M	16·50
Walnut (T.L.) W.	3rd cut	M	21·60
Lane	2nd cut	O	16·80
Lane	3rd cut	O	18·80

PROTEIN ANALYSIS OF CROPS 1963

Percentage as received

CROP AND SECTION		*Total protein*
Oats		
7 Acre (winter)	M	12·00
7 Acre (spring)	M	12·80
L. Wassex S. (winter)	O	13·10
L. Wassex S. (spring)	O	13·20
Garden (spring)	O	11·80
Barley		
Chestnut E.	S	8·60
Little	S	9·50
Road	S	8·60
New Bells Long	M	11·60
First Break	O	9·20
Wheat		
Further Break	S	13·50
Big Middle	M	13·30
L. Wassex N.	O	13·00
Garden	O	12·60
Beans		
Chestnut E.	S	22·50
Stackyard	M	26·40
Nappers	O	24·60

Percentage as received

CROP AND SECTION		*Total protein*
Peas		
Stackyard	M	24·50
Stable (own seed)	O	23·20
Stable (bought seed)	O	23·60
Silage—Lucerne		
Walnut W.	M	1·92
Walnut E.	M	1·92
Lane	O	2·41
Silage—Oats and Peas		
Pond	M	1·64
Bushey Close W.	O	2·36
Silage—Ley		
Sawpit	M	1·84
Wood	O	2·30
Hay—Lucerne		
Walnut W.	M	13·10
Walnut E.	M	11·70
Lane	O	12·70
Hay—Ley		
Big South	M	9·46
Bushey East	O	7·70
Mid. Wassex	O	7·54
P.P. Hay		
Long Meadow	M	7·97
Dove House	O	16·70

APPENDIX 7

Haughley Soils: Available Magnesium 1954–65

Year	*Mg. per cent as analysed*				*As percentage of farm average*		
	Farm ave.[1]	*O*	*M*	*S*	*O*	*M*	*S*
1954	13·1	15·3	11·2	12·6	116	86	96
1955	10·5	11·3	9·8	10·3	108	93	98
1956	10·5	11·4	10·0	9·7	109	95	92
1957	21·4	21·7	22·7	18·8	101	106	88
1958							
1959	23·8	24·6	24·6	20·6	103	103	87
1960	23·4	24·0	23·7	21·8	102	100	93
1961	24·5	23·5	25·8	24.0	96	105	98
1962	27·0	31·0	26·3	21·1	114	97	78
1963	36·1	37·1	34·1	37·7	103	94	104
1964[2]	37·3	36·3	37·7	39·3	97	101	105
1965[3]	40·5	41·0	40·0	47·7	101	99	100

[1] Note: Average of all farm crops except lucerne, quarterly analyses.
[2] Average of two readings per field.
[3] Average of one reading per field.

APPENDIX 8

Elaboration of Proposed Future Research

by Dr. R. F. MILTON
[1962]

ANIMAL FEEDING TESTS

Feeding experiments of a rather simple pilot type have been carried out on wheat, one by Rothamsted, and one by the Royal Free Hospital Medical School, reported to the Nutritional Society in 1958. Both produced inconclusive results perhaps because only one particular foodstuff was fed.

We would suggest that a scheme for the continuous feeding of small animals be evolved, involving the turnover of a number of generations, but with a diet which is both nutritionally balanced according to orthodox standards and yet at the same time of an overall comparison of the food value of crops grown on the various sections at Haughley.

The scheme should be elaborated after careful consideration by geneticists, and animal nutritionists, and should be devised so that as far as possible within the scope of the produce available from Haughley, the diet should be unmonotonous and to some extent auto-selective.

It is suggested that a number of pairs of small animals be grouped, and the feeding experiments be carried out with them and selected numbers of their progeny for at least ten generations.

In view of the failures associated with the above-mentioned pilot experiments, careful planning and frequent re-appraisal would be required of any scheme which might be suggested. This experiment should preferably be carried out in an animal laboratory fully equipped and experienced in such work. It is not considered that this could be done within the scope of our own organization.

PROTEIN, AMINO ACIDS AND NUCLEIC ACIDS IN CROPS AND PRODUCE

Differential analysis of the protein substances in crops could be a valuable indication of one aspect of nutritional value. Work on protein content of plants has mainly been concerned with total values based on destructive analysis for nitrogen content. Very little so far, has been done to evaluate the proteins as entities with a view to assessing possible nutritional differences in crops grown under differing conditions, although much work has been done to identify the type of protein which occurs in specific plant material.

A plan has been elaborated involving electrophoresis technique for the separation and determination of the protein material which occurs in plants. Solvents of various types which extract the protein material as little changed as possible would be investigated and the resultant solution eventually analysed by the electrophoretic technique. It is visualized that this analysis be carried out on the crops and herbage at various times in the growth phase, and also on the animal products from both stockbearing sections.

A PRELIMINARY SEARCH FOR FURTHER ACCESSORY FOOD FACTORS

We have been for the past few years carrying out regular analyses of the crops and produce from the sections. These analyses have included the usual items of recognizable food value, as well as mineral content (including trace elements) and also certain of the vitamins. There is more than inferential evidence that in addition to these factors—and also the differential proteins mentioned above—certain substances may be present in foodstuffs which affect palatability, digestibility and assimilability, and which could represent an important part of the nutritive value which would be included under the heading of quality.

In view of the fact that these factors are at present undefinable, any attempt to show such presence is mainly exploratory and speculative. Nevertheless, by making use of more recent developments in modern instrumental technique, any differences which might be present in foodstuffs grown under differing fertilizer and cultural conditions, might be highlighted.

A suggested line for such research of a preliminary nature is to 'screen' the products of the section by such instrumental technique as is available. The possible groups of substances involved would include:

(a) Essential oils.
(b) Water-extractable substances.
(c) Alcohol extractable substances.
(d) Fatty substances other than (a).
(e) Natural bases.
(f) Glucosides.

The suggested screening investigation should include the submission of the produce from all the sections to the above extraction techniques, and then examination by:

(1) Paper (two-way) and thin-layer chromatography techniques.
(2) Column chromatography including ion-exchange.
(3) Liquid-gas chromatography.
(4) Spectrophotometric pattern: (a) visible, (b) ultra-violet, (c) infra-red.

(1) PAPER CHROMATOGRAPHY

The value of this technique for the separation of acid and basic substances in biological fluids (or extracts) has been demonstrated by many workers, and methods are readily available for the proposed investigation. In addition, the method can be applied to the separation of 'larger molecules', e.g. plant xanthophylls and other pigments.

(2) COLUMN CHROMATOGRAPHY

The use of porous particulate including ion-exchange resins, has been recently resuscitated. This technique is suitable also for the separation of larger molecules. In addition, the so-called liquid chromatogram technique of Reichstein can be applied for the separation of pigments and naturally occurring organic complexes, as well as basic substances.

(3) GAS CHROMATOGRAPHY

Development of this technique has allowed methods which may be applied in the field under consideration to be used in the separation and identification of (a) essential oils, (b) 'flavours', (c) esters, (d) ethylenic unsaturated compounds, (e) fatty acids, (f) aromatic compounds. New applications are being described regularly, and it is certain that a thorough investigation using this technique applied to the extracts mentioned above, would reveal characteristic differences in the crops from the three sections, if they exist.

In addition, this technique may be used for the determination of minute amounts of insecticide residues. The chlorinated insecticides may be determined in concentrations as low as 0·02 ppm using a recently described gas-chromatographic technique.

(4) SPECTROPHOTOMETRIC PATTERN

(a) Visible, and (b) Ultra-Violet

Biochemical applications of absorption spectra in the visible and ultra-violet regions have (amongst others) allowed for much of the recent information concerning structure and function of certain enzymes, including co-enzyme A, and to an understanding of the mechanism of plant hormones. Standardization of preparative technique can allow the establishment of defined graph patterns which are characteristic of the molecular substance and electronic activity of the components.

The use of this instrumental approach for identifying the possibly different substances which might be resultant in produce from the different sections, becomes obvious, as firstly we would obtain a difference in the characteristic graph pattern, which becomes clearer with subsequent 'clean-up' procedures, and which may then be related to molecular composition and configuration.[1]

(c) Infra-Red Spectrophotometry

During recent years analysis by infra-red absorption has achieved a high degree of accuracy and has been widely applied. The analyses are empirical in the sense that they require calibration of the apparatus with the components of the mixture to be analysed.

As well as known specific compounds, infra-red spectroscopy may be used to determine those chemical groupings which give characteristic absorptions. The systematic study of absorption intensities has been the subject of considerable investigation during recent years. As a consequence it is now possible to interpret absorption graphs with more certainty and to obtain definite information concerning the chemical configuration of unknown products. Infra-red spectroscopy may, therefore, be applied to the substances under review in the same way as with the ultra-violet technique.

[1] See review by R. A. Morton, *Proceedings of the International Symposium of Microchemistry*, Pergamon, Oxford, 1959, p. 355.

These approaches to the investigation of the unknown substances in plant and foodstuff extracts should, of course, be carried out in conjunction with paper, column, and liquid-gas chromatographic partitions, which appropriately allows of group and individual substance separation and concentration—this being a necessary prerequisite to identification and possible determination.

An investigational plan of the above kind requires considerable care in outline and also in detail. The chosen method of extraction must be such that the material under study is opened up without undue alteration to the natural products which it contains, as anomalous results could be obtained due to the production of artifacts. A more reasoned programme would, of necessity, emerge subsequent to a preliminary survey. The approach is essentially experimental, and the investigator would be always on the look-out for leads which would channel further avenues of research.

Considerable use of the above tools has been made in many fields of biochemical research during the past decade, and there is much experience available by the specialized workers concerned. It should not be too difficult for such a worker after some initial experiments to formulate a precise research plan which might well lead to the uncovering of useful information. It cannot be emphasized too much, however, that the approach is initially limited by lack of knowledge of the precise object, and there may be difficulty in eliciting interest from an adequately trained worker because of the apparent nebulous nature of the commission.

THE PHYSICAL AND CHEMICAL BASIS OF PALATABILITY AND ITS SIGNIFICANCE TO NUTRITION

Experimental work on palatability is difficult to design. Any such work would be linked to the laboratory investigations being carried out under the search for qualitative differences.

VARIATION IN FLORAL COMPOSITION OF THE RUMEN OF HERBIVOROUS ANIMALS

Herbivorous animal digestion differs from those of carnivores and humans in that it substantially relies on bacterial decomposition of ingested foodstuffs for breakdown to simpler units capable of passing through the intestinal membrane into the body proper. The breakdown of cellulose to assimilable units is a very important feature of such a process, as is also degradation of plant proteins to amino acids.

The available information concerning the effect of various foodstuffs on the rumen flora is rather limited and little is known concerning the economy of the processes involved. Digestibility trials are carried out in various stations whereby known weights of ingested analysed food are compared with the amounts excreted unassimilated in the faeces. We have reason to believe from our work at Haughley that the milking cows on the Organic section appear to require less foodstuffs to maintain condition and to produce more milk than the corresponding animals on the Mixed section. It would be of interest to investigate whether this was connected with the assimilability coefficient of the food in question, using the established methods as mentioned above. An extension of this idea would be to compare the bacterial population of the various stomachs of the beasts as well as that of the faeces, to ascertain how this becomes modified according to variation in the quality of the food intake.

Such work would be invaluable in furthering a general understanding of animal nutrition, and could be carried out at any of the Agricultural Research Stations with facilities for conducting such tests.

APPENDIX 9

Huntingdon Feeding Experiment

In the second half of 1963, thanks to the generosity of the Rothley Trust, we had the opportunity to do small pilot experiments at the Huntingdon Laboratories of the Huntingdon Research Unit. Groups of standardized laboratory rats and guinea pigs were fed from weaning to maturity—about ten weeks—on diets consisting of Haughley produce: one group being fed on produce from the Mixed section, and the other on produce from the Organic section. The object of this experiment was to see whether it was possible to show—as a result of differing growth curves in the two groups of animals—whether any nutritional differences can be demonstrated. In point of fact, however, the results were very inconclusive, largely as a result of utilizing a diet which was quite unnatural for these animals. It would seem, being wise after the event, that we too quickly consented to submitting a diet which represented a compromise between what is fed to the animals at Haughley and what would be acceptable to the laboratory animals. In point of fact, the growth curves which we obtained reflected the differences in the ability of the individual animals to adapt themselves to the diet, rather than nutritional differences in the diets themselves. All that we have learned from this pilot experiment is that there are many technical difficulties to be overcome before we can launch out on a full-scale ten generation experiment using small animals to compare possible differences in the sectional produce. This must apply to both the rat and guinea pig experiments which were carried out simultaneously.

The third experiment was more successful and instructive. The Huntingdon Laboratories possess two fistulated heifers, i.e. animals with tubes fixed into their rumen, so that samples of the food ingested can be abstracted after varying time intervals. These animals are perfectly contented and suffer no pain or discomfort as a result of this permanent fistula. The two animals were fed for four months on Haughley produce: Heifer (A), Organic produce first and second months, Mixed produce third and fourth months; Heifer (B), Mixed produce first and second months, Organic produce third and fourth months. During the second and fourth months, samples of the partly digested food were taken from the animals for analysis. The investigation included examination of the bacterial flora, and fatty acid composition. These latter analyses were carried out using the gas chromatograph technique in the Welbeck Laboratories, and some very marked differences have been found. During the last two or three years, work on rumen digestion has been carried out in some of the Agricultural Research Stations in this country, and it has been shown that different diets produce different ratios of the simple organic acids, e.g., acetic, propionic, and butyric, and that the synthesis of butterfat in the cow's udder is markedly affected accordingly.

We are not yet in a position to fully interpret the results we obtained with the fistulated heifers, but these findings may help to bridge the gap in our knowledge as to why the Organic produce appears to result in higher milk yields.

APPENDIX 10

The Suggestion of Genetic Explanations for Higher Milk Yields in the Organic Herd Examined

The foundation animals consisted of thirty-four heifers bought, in 1947 and 1948, mostly in pairs of half-sisters by the same sire and out of cows of similar records. Some were in groups of four half-sisters by the same sire. In the first case one of each pair was allocated to each herd, and in the second case two of each. They were 'picked' alternately by two experts, one acting for each herd. Eight sires were involved thus each herd started with eight 'families'.

By 1962 two had dropped out of both herds—of two others one had dropped out of the Organic herd and one out of the Mixed—that is to say by the direct female line. The descendants of the two original sires in question, in the male line both produced a number of half-sisters that were still represented in both herds.

Of the original foundation half-sisters (all of which had left the farm by 1962) six of those allocated to the Mixed herd gave over 8,000 lb. in their best lactation, while only three of those allocated to the Organic herd did so. It seems unlikely, therefore, that later trends could have resulted from the original allocations.

Home-breds on both herds were always sired by the same bull.

The following tables give all the completed lactations of every heifer calf bred on the farm since the beginning of the Experiment, from their first calving, up to, and including the year 1962, with the exception of those used as nurse cows whose yields could not therefore, be recorded.

The animals belonging to each section are shown side by side, in groups of daughters by the same bull.

The original family of each animal is indicated by the code letters in the first column, which stand for the sire of the original foundation cow.

The obvious comment on the tables is that the general average of yields is low. Herd improvement brought about solely through home breeding and culling (which a commercial farmer would not do) is a slow process, and it should once more be emphasized that it is the deliberate policy not to employ any forcing methods, since one of the factors being investigated is the true long-term production capacity of the land of the two sections, in terms of end products. Thus no bought concentrates are used, and no special 'steaming-up' is practised. Pre-calvers never get more than 4 lb. per head per day of homegrown grain feed (often less), and this for only the last four weeks before calving. If transferred to a herd where the usual commercial forcing is practised there is some evidence to indicate that average yields would probably be increased by over 200 gallons per head. Nevertheless, that the Haughley system of management and feeding rates is capable of producing yields well up to the required standard of a commercial Guernsey herd that uses bought cattle cake, is shown by the few lactations exceeding 1,000 gallons.

An average of 700 for heifers and 850 for older cows is the target aimed at. The position so far is that of the thirty-four Organic home-breds there listed, five gave over 700 gallons, and four over 900 gallons.

To seek out the influence of the original foundation family, the homebreds' yields have also been arranged according to their descent from the foundation cow, and as before by the number of the lactation.

DAUGHTERS OF HOLMBURY ROSELAD 10TH

ORGANIC

Family	Born	Name		Lactations
M.A.	1950	Misty Maid	GB	8,640, 6,184, 7,266, 9,914, 8,641, 7,540, 7,715, 8,986, 7,713
T.L.	1951	Pyramid		6,228, 6,213, 6,836, 6,790, 6,206
T.L.	1951	Eve		6,219
B.R.	1951	Celandine		5,797
T.L.	1951	Begonia		8,245, 6,975, 7,651
B.R.	1952	Susan 2nd		3,387
B.R.	1952	Alice 2nd		6,342
T.J.	1952	Haughley Poppy	P	3,247, 2,422
T.L.	1952	Pyrethrum 2nd		5,100, 3,329
B.R.	1955	Lupin 4th		7,131, 8,050

One son of Holmbury Roselad 10th was kept and reared as a stock bull for use on both sections. He was named Haughley Roselad and was out of Rex's Bluebell of Newdigate, the foundation cow, by Tackley Rex 6th (allocated to the Organic section)

MIXED

Family	Born	Name		Lactations
M.A.	1950	Penny		4,146, 6,005, 4,604, 4,749
T.J.	1951	Haughley Bounty	P	5,925, 5,784, 7,138, 6,561, 5,993, 8,104, 5,255
T.L.	1951	Budleia		6,199, 6,329
D.V.	1952	Beauty 2nd		6,458, 5,919, 6292, 5,772
V.L.	1952	Flo	GB	6,804, 6,132, 5,024, 4,904
T.L.	1954	Sally 3rd		4,912, 3,195
T.J.	1957	Haughley Posy 3rd	P	6,098

DAUGHTERS OF TORIAN ROSA'S RENOWN

ORGANIC

Family	*Born*	*Name*		*Lactations*
D.V.	1952	April 2nd		6,667, 5,255, 4,459, 6,710, 3,842
M.A.	1952	Frosty Maid		8,560, 7,448, 9,920, 9,457, 7,127, 10,174, 9,065
B.R.	1953	Lupin 2nd		5,601, 4,259
V.L.	1953	Clover 2nd		6,330, 6,073, 4,736
M.A.	1953	Misty Maid 2nd		4,329, 1,817
T.L.	1954	Primrose 2nd		4,750
T.L.	1954	Begonia 3rd		5,125
M.A.	1955	Windy Maid	GB	10,168, 6,934, 9,784, 9,943, 9,449
V.L.	1955	Myrtle 3rd	GA	7,654, 4,684, 7,000, 6,764, 7,221
T.J.	1957	[1]Haughley Bouquet	P	8,887, 5,088, 6,375
M.A.	1957	Napper's Maid		7,329, 4,828

MIXED

Family	*Born*	*Name*		*Lactations*
V.L.	1952	Chloe 2nd		6,895, 4,883, 9,949
D.V.	1953	Beauty 3rd	GB	7,504, 6,120, 6,696, 7,157
T.R.	1953	Haughley Syringa	P	5,705, 3,797
T.L.	1953	Sally 2nd		4,690, 3,555
T.J.	1953	[1]Haughley Posy	P	6,344, 5,239, 5,131
T.J.	1953	Haughley Bounty 2nd	P	7,222, 3,124, 5,995
V.L.	1953	Flo		6,804, 6,465, 5,024, 4,904
V.L.	1954	Chloe 3rd	GB	7,669, 5,360, 6,562, 5,310
T.J.	1954	Haughley Bounty 3rd	P	7,315, 6,709, 4,746
M.A.	1954	Penny 2nd		5,632
D.V.	1954	Beauty 4th		7,417
D.V.	1954	Beauty 5th		7,035, 4,262
T.L.	1954	Polyanthus 3rd		5,045

[1] Full sisters.

DAUGHTERS OF HAUGHLEY ROSELAD

(Thus also all grand-daughters of Holmbury Roselad 10th on Sire side)

ORGANIC

Family	*Born*	*Name*		*Lactations*
M.A.	1955	Misty Maid 3rd	GC	6,993, 8,325, 6,420, 9,494, 6,777
T.L.	1955	Pyramid 2nd	GA	7,636, 6,025
T.L.	1955	Pyrethrum 4th		5,377, 5,310
T.L.	1955	Primula 4th		6,617
B.R.	1955	Lupin 4th		7,131, 8,423
T.L.	1956	Pyramid 3rd	GA	6,637, 4,061
M.A.	1957	August Maid	GB	6,961, 7,201, 6,984

On Dam's side, Misty Maid 3rd, Pyramid 2nd and Pyramid 3rd, are double cross with Holmbury Roselad 10th. Lupin 4th and Primula 3rd are granddaughters of Torian Rosa's Renown, and August Maid and Pyrethrum 4th are daughters of foundation cows

MIXED

Family	*Born*	*Name*		*Lactations*
T.L.	1954	Sally 3rd		4,912, 3,195
D.V.	1955	Bluebell		(Suckled), 6,248, 6,250, 6,740
T.L.	1955	Sally 4th		6,227, 6,354, 5,923, 7,509, 6,584
D.V.	1956	Beauty 6th		6,297, 5,419
T.R.	1956	Haughley Syringa 2nd	P	6,157, 4,306
V.L.	1956	Flo 2nd		6,090
T.L.	1957	Sally 5th		5,140, 5,780, 5,355
V.L.	1957	Chloe 4th		6,378, 6,757, 5,892
V.L.	1957	Flo 3rd		7,746, 6,123, 2,310
M.A.	1957	Penny 3rd	P	5,879
T.J.	1957	Haughley Posy 3rd		6,098

On Dam's side, all the above (except Beauty 6th who is a double cross with Holmbury Roselad) are grand-daughters of Torian Rosa's Renown

DAUGHTERS OF HOMEWOOD ROBERT 22ND

ORGANIC					MIXED				
Family	*Born*	*Name*	*Lactations*	*Dam's sire*	*Family*	*Born*	*Name*	*Lactations*	*Dam's sire*
M.A.	1958	Nappers Maid 2nd	5,533	Haughley Roselad	D.V.	1958	Bluebell 2nd	4,464, 4,518	Haughley Roselad
T.L.	1958	Pyramid 4th	3,964	Haughley Roselad	T.L.	1958	Sarah	5,259, 4,686	Haughley Roselad
M.A.	1959	Freezing Maid	4,867, 3,627	Torian Rosa's Renown	V.L.	1958	Chloe 5th	5,742, 5,112	Torian Rosa's Renown
					T.J.	1958	Haughley Pretty P	5,292, 4,005	Torian Rosa's Renown
					D.V.	1958	Beauty 7th	5,932, 4,489	Holmbury Roselad 10th
					T.J.	1958	Bounty 5th	4,482	Holmbury Roselad 10th

DAUGHTERS OF HOMEWOOD ROBERT 27TH

ORGANIC					MIXED				
Family	*Born*	*Name*	*Lactations*	*Dam's sire*	*Family*	*Born*	*Name*	*Lactations*	*Dam's sire*
M.A.	1959	Spring Maid	5,467	Holmbury Roselad 10th	T.J.	1959	Haughley Bounty 6th P	5,413	Holmbury Roselad 10th
B.R.	1959	Lupin 6th	1,843	Holmbury Roselad 10th	T.R.	1959	Haughley Syringa 3rd P	5,037	Haughley Roselad
M.A.	1959	December Maid	5,168	Torian Rosa's Renown	T.J.	1959	Haughley Poppy P	3,888, 2,753	Torian Rosa's Renown

AVERAGE LACTATION YIELDS

(number of cows in brackets)

(BY SIRES)

Daughters of		*Holmbury Roselad 10th*	*Torian Rosas Renown*	*Haughley Roselad*	*Homewood Robert 22nd*	*Homewood Robert 27th*
Lactation no.						
I	O	6,034 (10)	6,855 (11)	6,765 (7)	4,788 (3)	4,159 (3)
	M	5,792 (7)	6,560 (13)	6,092 (10)	5,195 (6)	4,779 (3)
II	O	5,529 (6)	5,154 (9)	6,558 (4)	3,627 (1)	
	M	5,561 (6)	4,951 (10)	5,523 (8)	4,562 (5)	2,753 (1)
III	O	7,251 (3)	7,046 (6)	6,702 (2)		
	M	5,764 (4)	6,300 (7)	5,146 (5)		
IV	O	8,352 (2)	8,218 (4)	9,494 (1)		
	M	5,496 (4)	5,790 (3)	7,124 (2)		
V	O	7,424 (2)	6,910 (4)	6,777 (1)		
	M	5,993 (1)		6,584 (1)		
VI	O	7,540 (1)	10,174 (1)			
	M	8,104 (1)				
VII	O	7,715 (1)	9,065 (1)			
	M	5,255 (1)				
VIII	O	8,986 (1)				
	M					
IX	O	7,713 (1)				
Average all lactations	O	6,622 (27)	6,773 (36)	6,843 (17)	4,498 (4)	4,159 (3)
	M	5,763 (24)	5,775 (34)	5,833 (26)	4,907 (11)	4,272 (4)
	Mean	6,217 (51)	6,288 (70)	6,233 (43)	4,794 (15)	4,209 (7)

		Organic	*Mixed*	*Mean*
Average all bulls	I	6,174 (34)	5,955 (39)	6,064 (73)
	II	5,570 (22)	5,087 (30)	5,291 (52)
	III	7,039 (11)	5,806 (16)	6,307 (27)
	IV	8,439 (7)	5,956 (9)	7,042 (16)
	V	7,038 (7)	6,288 (2)	6,871 (9)
	VI	8,857 (2)	8,104 (1)	8,606 (3)
	VII	8,390 (2)	5,255 (1)	7,345 (1)
	VIII	8,986 (1)		
	IX	7,713 (1)		
	Mean	6,545 (87)	5,630 (99)	6,058 (186)

Difference between the two herds by lactations. (The advantage is to the Organic herd in each case.)

Lactation	*Difference, lb. milk per lactation*	*Difference as percentage of mean of two herds*
I	219	3·6
II	483	9·1
III	1,233	19·5
IV and subsequent	1,348	18·5

AVERAGE LACTATION YEILDS BY FAMILIES

(number of cows in brackets)

Family		*Boconnoc Ruler*	*Trewithin Lodestone*	*Violet Lad*	*Mithian Armistice*	*Tackley Rex*	*Dinahs Victor*	*Trewithin Jupiter*
Lactation								
I	O	5,319 (7)	5,991 (11)	6,992 (2)	6,729 (11)	—	6,667 (1)	6,067 (2)
	M	—	5,298 (8)	6,766 (8)	5,219 (3)	5,633 (3)	6,444 (7)	5,807 (10)
II	O	6,911 (3)	5,319 (6)	5,378 (2)	5,796 (8)	—	5,255 (1)	3,755 (2)
	M	—	4,728 (7)	5,833 (7)	6,005 (1)	4,052 (2)	5,282 (7)	4,602 (6)
III	O	—	7,244 (2)	5,868 (2)	8,075 (5)	—	4,459 (1)	6,375 (1)
	M	—	5,639 (2)	5,794 (6)	4,604 (1)	—	6,413 (3)	5,752 (4)
IV	O	—	6,790 (1)	6,764 (1)	9,702 (4)	—	6,710 (1)	—
	M	—	7,509 (1)	5,039 (3)	4,749 (1)	—	6,556 (3)	6,561 (1)
V	O	—	6,206 (1)	7,221 (1)	7,999 (4)	—	3,842 (1)	—
	M	—	6,584 (1)	—	—	—	—	5,993 (1)
VI	O	—	—	—	8,857 (2)	—	—	—
	M	—	—	—	—	—	—	8,104 (1)
VII	O	—	—	—	8,390 (2)	—	—	—
	M	—	—	—	—	—	—	5,255 (1)
VIII	O	—	—	—	8,986 (1)	—	—	—
IX	O	—	—	—	7,713 (1)	—	—	—

It has been suggested in some quarters that the chance allocation to the Organic herd, of a heifer that turned out to be the best foundation cow (Spotless by Mithian Armistice), is solely responsible for the higher average on that section. In order to find out if this was so, the table on p. 363 has been repeated, omitting the Mithian Armistice family.

The result is as follows:

AVERAGE LACTATIONS OMITTING MITHIAN ARMISTICE FAMILY

(According to Sires)

Daughters of		*Holmbury Roselad 10th*	*Torian Rosa's Renown*	*Haughley Roselad*	*Homewood Robert 22nd*	*Homewood Robert 27th*	*Average*
Lactation							
I	O	5,744 (9)	6,431 (7)	6,680 (5)	3,964 (1)	1,843 (1)	5,909 (23)
	M	6,066 (6)	6,637 (12)	6,116 (9)	5,195 (6)	4,779 (3)	6,017 (36)
II	O	5,398 (5)	5,072 (5)	5,955 (4)	—	—	5,440 (14)
	M	5,472 (5)	4,951 (10)	5,523 (8)	4,562 (5)	2,753 (1)	5,056 (29)
III	O	7,244 (2)	5,642 (4)	—	—	—	6,176 (6)
	M	6,151 (3)	6,300 (7)	5,146 (5)	—	—	3,886 (15)
IV	O	6,790 (1)	6,737 (2)	—	—	—	6,755 (3)
	M	5,746 (3)	5,790 (3)	7,124 (2)	—	—	6,109 (8)
V	O	6,206 (1)	5,532 (2)	—	—	—	5,756 (3)
	M	5,993 (1)	—	6,584 (1)	—	—	6,288 (2)
VI	O	—	—	—	—	—	—
	M	8,104 (1)	—	—	—	—	—
VII	M	5,255 (1)	—	—	—	—	—

This shows that the presence of Spotless's descendants (all the Maids) have had a considerable influence on resulting averages, but also that the overall picture would not be reversed by their removal. This is perhaps easier to see in graph form, as under.

GRAPH 8

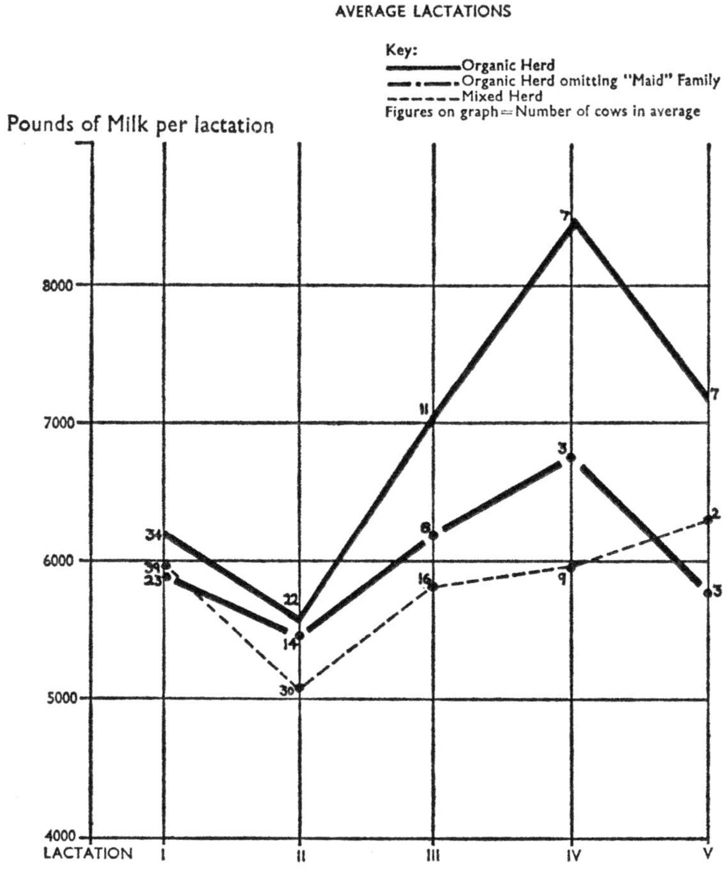

APPENDIX II

Dr. F. M. Pottinger's Cat-feeding Experiments
Dr. W. Albrecht on Role of Organic Matter in Soil

References: F. M. POTTINGER, Jnr., MD. 'Feeding experiment on cats', *American Journal of Orthodontics and Oral Surgery*, XXXII, 8, August 1946.

This experiment extended over ten years and involved 900 animals. The main purpose was a comparison between cooked and raw food, though there were various subdivisions using different combinations, such, for example, as groups of cats fed on raw meat with pasteurized milk, and others on cooked meat with raw milk. The animals who received an all-raw-food diet, both milk and meat, remained healthy and bred normal healthy litters from generation to generation, while all those of which cooked food formed the major portion of the diet, whether this was meat or milk, became progressively degenerate through succeeding generations. For example, 25 per cent of abortion occurred in the first generation and 70 per cent in the second. The animals also fell prey to a varied range of diseases, all listed in the report, and in many cases by the third generation the kittens had become so degenerate that they failed to survive for six months. A further experiment with different kinds of milk produced the same result. The raw-milk-fed cats remained healthy and bred normally from generation to generation, while all those fed on other forms of milk suffered from increasing degrees of sickness, degeneration and skeletal malformation in this order—pasteurized milk, evaporated milk, and sweetened condensed milk. In later experiments cats whose general metabolism had been deranged by the cooked food were returned to a raw food diet. Complete regeneration, where it was not too late to achieve this, took four generations.

The report ends with this extremely significant statement, 'The principles of growth and development are easily altered by heat and oxidation which kill living cells at every stage of the life process from the soil, through the plant and through the animal. Change is not only shown in the immediate generation but as a germ plasm injury *which manifests itself in subsequent generations of plants and animals.*'

On the role of organic matter in soil, Dr. Albrecht[1] confirmed that plants can and do take up organic compounds, and that this is certainly an important factor in both plant and animal nutrition.

'We have been so schooled in plant nutrition to consider only inorganic ions that we have arrived at the belief that therefore plants do not use organic compounds. . . . We are slowly coming to consider the fact that plant roots absorb organic compounds directly from the soil for metabolic services in the plant's synthesis of its own organic substances, possibly proteins.'

He gave me much evidence to support this view. Water extracts of organic matter in soils serve as growth hormones in the laboratory for test plants. Black currants and potatoes have been demonstrated to take up benzinehexachlorate, and plant roots can take up indole. In connection with this last, he threw a new light on Dr. Pottinger's famous cat experiments and, incidentally, on the biosynthesis by plants of amino acids. Members familiar with these experiments will remember that, in one, cats were fed for two years in four different pens. All feed components for these cats were constant, except the milk. Four kinds of milk were used, (a) sweetened condensed, (b) evaporated, (c) pasteurized, (d) raw. The health of the cats was in inverse ratio to that order. At the close of the experiment, volunteer weeds grew up in the pens, which had as their flooring material clean, quartz sand. The vigour of the weed growth followed the same pattern as the health of the cats.

'All the evaporated and heated milks coming by way of the cat dung apparently did not put into the sand enough fertility even to invite weed growth. Raw milk had put so much back, even after feeding the cats better, that the weed growth filled the pens completely.'

As a further test, the weeds were dug in and Michigan White dwarf beans planted in all four pens. They, too, followed the same pattern as the weeds; but in this case, even growth habit was changed, for in the raw milk pen the beans ceased to be dwarf and climbed the wire 6 ft. high. These beans were left to ripen and the seed harvested. All the seed except that from the raw milk pen smelt of cat excreta.

[1] William Albrecht, PhD. Late chairman of the Department of Soils, College of Agriculture, University of Missouri, USA.

This odour is caused by the common faecal excretions indole and scatole, which are ring compounds unbroken by digestion. Indolacetic acid is the plant hormone giving pronounced growth of roots and shoots. Indole becomes this hormone by addition to the indole ring. With little further change by addition, it becomes tryptophane, the frequently deficient, but required, amino acid. So, having indole in the cat dung, the suggestion of a hormone to change dwarf beans into pole beans, the presence of indole odour in all the ripe seed except the 'pole' beans in the raw milk pen, Dr. Albrecht's suggested interpretation is that the beans took up the indole in unbroken form, but in the case of those growing on the dung of the raw-milk-fed cats it was converted into indolacetic acid, and possibly tryptophane, as part of the bean protein. He further suggested that this may very well be the normal route of travel of the organic compounds in the cycle from soil to plant to animal and back to the soil. In this particular case, two animals were involved, the cow and the cat, and the effect of merely sterilizing the milk was in some way to upset the normal flow of these organic compounds in the cycle.

Dr. Albrecht believes the unerring instinct of animals for choosing food of high nutritional value to be an awareness of factors in this organic cycle, rather than of simple mineral ones. One experiment demonstrated this very clearly. Three plots of land were treated respectively with (a) lime, phosphate and potash, (b) lime and phosphate, (c) lime only. The crop rotation on each plot was a red clover ploughed down green, followed by corn (maize), followed by wheat. The three wheat crops were offered in feed hoppers simultaneously to pigs. The order of their choice was (a), (b), (c). Many workers would have been content to leave it at that—not so Albrecht. When sweet clover was substituted for red clover as a green manure crop for corn, followed by wheat, and choice of the three wheat crops was again offered to pigs, their order of preference became (b), (a), (c). In another experiment, using the same three soil treatments, corn was offered from the three plots, following sweet clover as green manure. The choice then was (c), (b), (a). When, however, the sweet clover was allowed to mature for seed and harvested, leaving the clover trash on the ground for the corn crop, the choice for that corn crop was once more (a), (b), (c).

From these results, Albrecht argues that the hogs were not voting for the ash content of the corn or wheat resulting from the application of the minerals, nor for bulk yield, but rather for some effects prompted by the nature of the organic material turned under as green manure.

In other words, 'mineral fertilizers are not of influence only as minerals, rather they are in control of what products the plants manufacture in consequence of the minerals' presence in the soil.' Many of his other experiments have also clearly demonstrated that an increase in vegetative growth is no criterion for assuming an increase in food production.

APPENDIX 12

Source Material

Report	*When published*	*Period covered*
The First Twenty-Five Years	June 1962	1938—1960–1
Annual Report	June 1963	1961–2
Annual Report	December 1964	1962–3

(The above include summaries of the analytical work, most of which are reproduced here in Appendices 2–7 and 9.)

Report	*When published*	*Period covered*
Farm Report	Not published	1964
Farm Report	Not published	1965
Farm Report	Not published	1966
Farm Report	Not published	1967
Farm Report	Not published	1968
Farm Report	Not published	1969
Summary of Analytical Work, (excluding Sheep and Poultry) compiled by J. F. Ward	Not published	1952–65
Addendum to above	Not published	1952–65

Details of all laboratory analyses carried out at Dr. Milton's laboratories (typescript).

Arrangements can be negotiated for all the above, to be made available for study on application to:

The General Secretary, Soil Association, Walnut Tree Manor, Haughley, Stowmarket, Suffolk IP14 3RS (Tel.: 044 970 235/6).

Bibliography and Recommended Reading

A number of works from the original Bibliography have been included on account of their particular importance as source material.

Barlow, K. E., *The Discipline of Peace* (Faber & Faber, 1942; Charles Knight, 1971).

Barrett, T. J., *Harnessing the Earthworm* (Faber & Faber, 1949).

Billington, F. H., *Compost* (Faber & Faber, 1942; 4th edn., 1956).

Bruce, Maye E., *From Vegetable Waste to Fertile Soil* (Faber & Faber, 1943).

——, *Common-Sense Compost Making* (Faber & Faber, 1946; rev. edn., 1967).

Carrel, A., *Man the Unknown* (Hamilton, 1935).

Carson, Rachel, *Silent Spring* (Hamilton, 1963; Penguin, 1970).

Christophersen, Erling, *Tristan da Cunha* (Cassell, 1940).

Darwin, Charles, *Darwin on Humus and the Earthworm: The Formation of Vegetable Mould*, with an introduction by Sir Albert Howard (Faber & Faber, 1945).

Dhar, Prof. N, *World Food Crisis and Land Fertility Improvement* (University of Calcutta, 1972).

Dreschler, C., 'Predacious Fungi', *Biological Review*, XVI, 1941.

Grant, Doris, *Your Daily Food* (Faber & Faber, 1973).

Hills, Hilda Cherry, *Living Dangerously* (Tom Stacey, 1973).

Hopkins, Donald, *Chemicals, Humus and the Soil* (Faber & Faber, 1945; rev. edn., 1957).

Howard, Sir Albert, *An Agricultural Testament* (OUP, 1940; Rodale, 1972).

——, *Farming and Gardening for Health or Disease* (Faber & Faber, 1945).

Jacks, G. V., and Whyte, R. O., *The Rape of the Earth* (Faber & Faber, 1939).

Jenks, J. E. F., *From the Ground Up* (Hollis & Carter, 1950).

——, *The Stuff Man's Made Of* (Faber & Faber, 1959).

Kervran, C. L., *Biological Transmutations* (English translation H. & E. Rosenauer, Crosby Lockwood, 1972; first published Le Courrier du Livre, 1966).
King, E. H., *Farmers of Forty Centuries* (Cape, 1926).
King, F. C., *The Compost Gardener* (Titus Wilson, 1943).
——, *Gardening with Compost* (Faber & Faber, 1940).
McCarrison, Sir Robert, *Journal of Medical Research*, XIV, 1926, p. 351.
—— and Sinclair, H. M., *Nutrition and Health* (Faber & Faber, new edn., 1962).
Northbourne, Lord, *Look to the Land* (Dent, 1940).
Osborn, Fairfield, *Our Plundered Planet* (Faber & Faber, 1948).
Pearse, I. H., *Health, of the Individual, of the Family, of Society* (booklet, Pioneer Health Centre, 1971).
—— and Crocker, L. H., *The Peckham Experiment: A Study of Living Structure of Society* (Allen & Unwin, 1943).
Pfeiffer, E., *The Earth's Face* (Faber & Faber, 1947).
—— *Soil Fertility, Renewal and Preservation* (Faber & Faber, 1947).
Picton, L. J., *Thoughts on Feeding* (Faber & Faber, 1946).
Portsmouth, Earl of, *Alternative to Death* (Faber & Faber, 1943).
——, *Famine in England* (Witherby, 1938).
Pottinger, F. M., Jnr., 'Feeding Experiments on Cats', *American Journal of Orthodontics and Oral Surgery*, XXXII, 8, August 1946.
Price, Weston, *Nutrition and Physical Degeneration* (The Price-Pottinger Foundation, New York, 1945; Heritage edn., 1972). Obtainable from Wholefood, 112 Baker Street, London W1.
Rayner, M. C., *Mycorrhiza* (Cambridge UP, 1927).
——, *Trees and Toadstools* (Faber & Faber, 1945).
——, 'The Mycorrhizal Habit in Crop Plants, with a Reference to Cotton', Empire Cotton Growing Corporation publication, reprinted from *Empire Cotton Growing Review*, XVI, 3.
—— and Neilson-Jones, W., *Problems in Tree Nutrition* (Faber & Faber, 1944).
Rowlands, T., *Biochemical Journal*, XXIV, 1, 1930.
Russel, Sir John, *World of the Soil* (Collins, 1957).
Schumacher, E. F., *Small is Beautiful* (Blond & Briggs, 1973).
Stapledon, Sir George, *Human Ecology*, ed. Robert Waller (Faber & Faber, 1964).
Sykes, Friend, *Food, Farming and the Future* (Faber & Faber, 1951).
——, *Humus and the Farmer* (Faber & Faber, 1946).

Tompkins, Peter, and Bird, Christopher, *The Secret Life of Plants* (Allen Lane, 1974).
Turner, Newman, *Fertility Farming* (Faber & Faber, 1951).
Waksman, S. A., *Humus* (Baillière, Tindall & Cox, 1937).
Westlake, Aubrey, *Pattern of Health* (Vincent Stuart, 1961).
Williamson, G. Scott, *Physician, Heal Thyself* (Faber & Faber, 1946).
Wrench, G. T., *Reconstruction by Way of the Soil* (Faber & Faber, 1946).
——, *The Restoration of the Peasantries* (Daniel, 1940).
——, *The Wheel of Health* (Daniel, 1938).

ORGANIZATIONS AFFILIATED TO, OR WITH THE SAME AIMS AS, THE SOIL ASSOCIATION, WITH THEIR PUBLICATIONS, IF ANY

The Organic Soil Association of South Africa, *Soil Sense* (quarterly).
The Soil Association of New Zealand, *Journal* of (quarterly).
Organic Farming & Gardening Society (Australia), *The Good Earth* (quarterly).
South Australian Group of the Soil Association.
The Henry Doubleday Research Association, *Newsletter*.
The McCarrison Society, *Reports* (periodical).
Land Fellowship (Canada), *Land Bulletin* (monthly).
L'Association Européenne d'Agriculture et d'Hygiène Biologiques (France), *Nature et Progrès* (monthly).
Organische Landbau (Germany), *Organische Landbau* (quarterly).
Svenska Biodynamiska Foreningen (Sweden).
The Organic Farmers' Association of Israel.
Organic Gardening & Farming (USA), *Organic Gardening and Farming* (monthly).
Natural Food Associates (USA), *Natural Food and Farming* (monthly).

Glossary

Aerobic Dependent upon air for life.
Amoeba A genus of microscopic Rhizopoda (see Rhizopod).
Anaerobic Capable of living in an atmosphere without oxygen.
Arbuscular Tufted, shaped like an arbuscle (i.e. a dwarf tree).
Bionomics Dynamic biology. Bionomy = the science of the laws of life, or of living structure.
Blepharitis Inflammation of the eyelid.
Caecum The blind gut. The first part of the large intestine which is prolonged into a cul-de-sac.
Cash Crops Any crop sold off the farm for cash, as opposed to one consumed on the farm by livestock.
Detritus Matter produced by the wearing away of exposed surfaces. An accumulation of disintegrated material or debris.
Diathesis A constitutional predisposition (inherited or acquired).
Ecology That branch of biology which deals with the relations of living organisms to their surroundings and each other.
Ecto Outside.
Edaphic Pertaining to soil.
Empirical Method based on result of observation and experiment, not on scientific theory.
Endo Within.
Endophyte A plant growing within an animal or another plant.
Endothelium The tissue lining blood vessels, etc.
Frumenty A country dish made of boiled whole wheat and milk.
Haustorium The sucker of a parasitic plant or endophyte which penetrates the tissue of the host.
Histology That branch of anatomy which is concerned with the cell structure of the body.
Hypae See Mycelium.
Inter Between.
Intra On the inside.
Lignin An organic substance which forms the characteristic part of

wood cells, and woody fibres, making the greater part of the weight of most dry wood.

Metabolism The sum of the chemical changes within the body, or within any single cell of the body, by which the protoplasm is either renewed or changed to perform special functions, or else disorganized and prepared for excretion.

Morphology The science of the outer form and internal structure (without regard to function) of animals and plants.

Motile Capable of deliberate motion.

Mycelium The vegetative system of fungi consists of filiform branched or unbranched cells called hypae. The hypae collectively form the mycelium.

Mycology The science of fungi.

Mycorrhiza Fungus mycelium which invests, and lives in association with, the roots of many plant species (see explanation, and definition, p. 73).

Nematode A class of worms, having a mouth and an alimentary canal and separate sexes. Usually parasitic. Examples: round worm, threadworm, eelworm.

Nodule A small knot or node in the stem or other part of the plant.

Otorrhœa A purulent discharge from the ear.

Optimal (See Optimum).

Optimum Best. — temperature. That temperature at which the metabolic processes are carried on with the greatest activity.

Pan An impervious layer in soil which interferes with the passage of air and water. Can occur at varying depths, and is due to a variety of causes.

Reticulum The network which pervades the substance of a cell and nucleus enclosing the softer portion of the protoplasm.

Rhizopod The lowest class of protozoa (primitive animal organisms).

Schlerotia A compact mass of hypae (see Mycelium) in a dormant state.

Silage Feed for cattle made by preserving fresh green fodder by compression in a silo.

Somatic Pertaining to the body, or material organism as opposed to soul, spirit or mind. — death = death of the body as a whole, i.e. all the cells of the body, as opposed to death in any of its parts.

Sporangiol Small spore case.

Sporophore The spore-bearing organ of a fungus. Example: mushroom, toadstool.

Symbiosis A living together of two organisms, each of which is necessary to the other.

Trophic Pertaining to nourishment, or nutrition.

Vascular Containing clearly defined vessels or ducts for the circulation of sap.

Vesicle Any small bladder-like structure, cavity, cell or the like, in a body.

Xerophthalmia Dryness of the eyes.

Index